My Father
Sergeant Rathel
American Soldier

***

Printed in the United States of America

First Printing, 2022

ISBN: 978-1-959640-02-8

Modern Warrior Publications: Division of Omega Group Consultants LLC.
490 County Road 39,

Headland, AL. 36345

modernwarriorproject.com

***

Greater Love Has No One Than This, That Someone Lay Down His Life For His Friends.
John 15:13 The Bible ESV

My Father Sergeant Rathel, American Soldier

***

About The Author

Randy Rathel lives in Dothan, Alabama is married to Denise Killebrew Rathel. He has two daughters and two bonus daughters. Together they have ten grandchildren. He is retired from the airline industry. They attend Harvest Church in Dothan, Alabama. Randy and his wife Denise love history, genealogy, and traveling to historic places. Randy's first book "Ruby's Memoires," about his Mother's life is loved by many.

***

I've known and considered Randy a good friend for nearly a decade now; so I was extremely pleased when he called me one day asking if I would consider helping him in the publication of his second book. I was aware of his success and the great reception of his first book reflecting on the memories of his mother titled "Ruby's Memoires" and eager to hear the topic of his second book. It's important to note here that Modern Warrior Publications, the DBA of our publishing company only deals with literature in certain topics. Those topics being Tactics, Survival, Preparation, and Fitness. So, while excited to hear about Randy's new work and his request for publishing I wasn't sure if we were the right fit. That was until I heard the details. Capturing the story of a dedicated Husband, Father and Patriot's military service is perfectly in our scope of publications. Not only that but being a retired Army Soldier with years of combat service of my own I was honored to have the opportunity in being part of paying homage to one who has went before me.

As you read through the pages to follow you will find the well-re- searched and detailed events of Sergeant Rathel's service during World War II. But, more than a simple recap of one's service rec- ord you will get a little history along the way, witness the im- portance of faith in his life, and perhaps best of all see the adora- tion of a son bleed through the pages as Randy seeks to honor his father's legacy in memoriam. I hope you the reader are uplifted and enjoy this literary dedication as much as I have.

Enjoy,
Shane
Blevins
Modern Warrior Publications a DBA
of Modern Warrior Project
Omega Group Consultants LLC

***

Table of Contents

Introduction

Honor and Grief-Emotions Felt at The Same Time

It was a beautiful Saturday afternoon as our family watched the three-volley rifle salute to Honor the life of my Father.

On May 13, 1989, my Father Sergeant Clarence Rathel was buried with full military honors.

With the emotions of Honor and grief, our family listened as the bugler played "Taps." Then the Honor Guard folded with precision The Flag draped across my Father's casket exactly thirteen times representing the original thirteen colonies of The United States of America; once the folding was completed no stripes were shown only the blue with stars. The Flag was now in the shape of the tricorn hat worn by the patriots of The American Revolution. The Flag was then presented to my Mother with the sincere condolences of our Nation.

I recently examined The Flag my Mother received that day, it had been given to me by her before she passed away. Inside The Flag were the 21 shell casings from the rifle salute. As I reflected upon the shells and that moment in time I remembered the verse of scripture that Reverend Dr. Burney H. Ensor used in the ceremony. Revelation 21:21; "*And the twelve gates were twelve pearls; every several gate was of one Pearl: and the street of the city was pure gold, as it were transparent glass." (NKJ)*

In his memorial message, Reverend Ensor explained that Pearls are formed by pain. The oyster experiences pain when a tiny grain of sand is trapped within its shell, it then secretes a fluid that covers the sand causing a Pearl to form. The Pearly Gates of Heaven are each One Pearl reminding us of the suffering of Christ and the pain humans experience in life. He also reminded us that in Heaven there is no pain.

As I thought of my Father, The Flag, and the memorial message I remembered the time we spent together during the last years of his life. During that time I brought him a small cassette tape recorder hoping he would record his life story, especially his military career. I did this with the desire that it would keep his mind off the constant pain he endured. My Father had never talked about his service in World War II, The Korean War, or his time in Germany during a crisis. The tapes he completed are fascinating and the world should hear his story.

A message to the reader from the author Randy Rathel

This book was written, to honor my Father's military service in the United States Army. It is dedicated to all who value Freedom, Liberty, and especially those who have, are, and will proudly serve our country in uniform. Inspired by the recordings of "Sergeant First Class Clarence Rathel's life," it tells the story of a farm boy from Alabama who like so many of his generation stood against the Goliath's of The Axis Powers in World War II and the North Koreans in the Korean War. For simplicity in writing, I will refer to his rank as Sergeant. With much diligence, I have condensed and made readable his story for clarity; adding background information where necessary. The italicized paragraphs give important historical information about the times and actual events Sergeant Rathel lived. It would be humanly impossible for me to describe the horrors and tragedy of the wars my Father experienced as described in his recordings. Now let's relive WWII and The Korean War from the eyes of Sergeant Rathel.

Thank you for reading, Randy Rathel

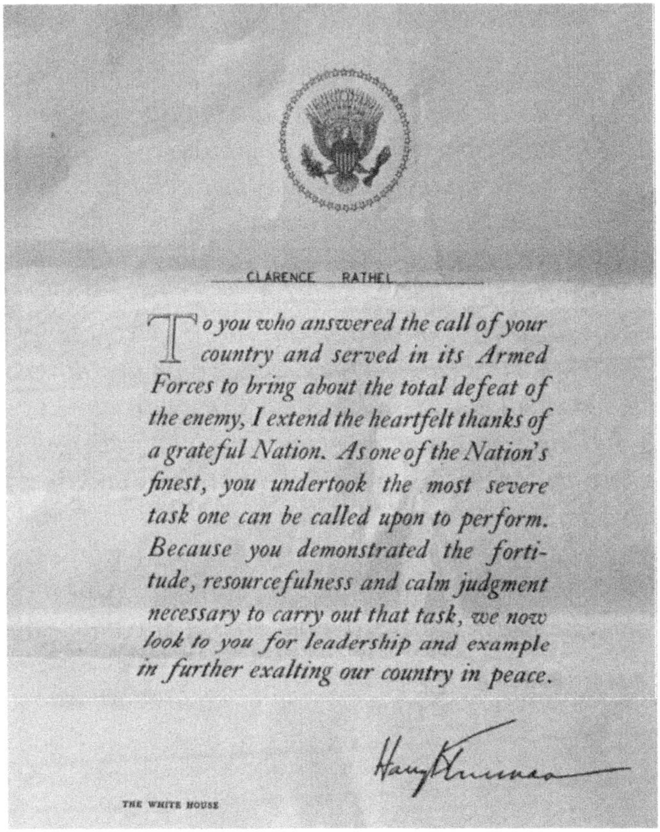

CLARENCE RATHEL

To you who answered the call of your country and served in its Armed Forces to bring about the total defeat of the enemy, I extend the heartfelt thanks of a grateful Nation. As one of the Nation's finest, you undertook the most severe task one can be called upon to perform. Because you demonstrated the fortitude, resourcefulness and calm judgment necessary to carry out that task, we now look to you for leadership and example in further exalting our country in peace.

Harry Truman

THE WHITE HOUSE

A letter of appreciation to Clarence Rathel for his service in World War II, signed by President Harry Truman

Picture: The Columbus Museum, Columbus, Georgia

***

Chapter One

Steamboats on The Chattahoochee

Sergeant Rathel – Speaking

My Father, Honeywell Rathel was a patriot who loved his country. He served as a soldier in The Spanish-American War which ended on August 13th, 1898. After returning home to my Mother and my siblings he purchased a blacksmith and stable company in Gordon, Alabama. He also operated a carriage business that transported passengers and cargo. Gordon in the early 1900s was a major port and shipping center for steamboats on The Chattahoochee River. The steamboats brought products and supplies to the three ports in Gordon. Cotton was still the major agricultural product in the south and the main cargo shipped to the harbor in Apalachicola, Florida for distribution to industrial locations.

Our family lived about three miles from Gordon and my Mother Lola Griffin Rathel managed our farm of 80 acres. With our large family, my older sisters and brothers worked the farm while my Father worked in town managing his other businesses.

In 1889 the Alabama Midland Railroad began service through Gordon which eventually brought an end to the steamboat era. During Gordon's transition to shipping by rail, our family slowly shifted direction becoming more focused on horses, cattle, and farming.

The Alabama Midland Depot in Ashford, Alabama, Established in 1889

Before I was born our family experienced the loss of three of my siblings. I had a brother named Albert who passed away at age six due to typhoid fever, a sister named Ruth who suddenly died when she was a baby of one, and a brother named Zema the family's oldest son tragically passed away at age twenty- three, the year before I was born.

My older sister said Zema was working with his horses one foggy November morning and collapsed with an aneurysm. Zema was engaged to a beautiful young lady and the wedding was planned for Christmas Eve. What was to be a joyful occasion became a time of sadness; for months after Zema passed away, our Mother would stand at the kitchen window, while looking at Zema's horse in the corral and weep.

I was born on November 19th, 1917 at midnight. Some in our family said I was born on the 20th. In the end, my Mother told me she made the decision that my birthday was on the 19th.

Not long after my birth, my Mother wanted to move to another place because she was still grieving the loss of her three children. I was told that somehow the home we were in was full of hurtful memories and she felt a different place would somehow take away the pain in her heart.

Zema Rathel, Gordon Cemetery, Gordon, Alabama

Soon our family moved to another farm. Our new home was a large log cabin that had two huge fireplaces. The farm was 160 acres. Our yard was beautiful, I remember the large oak trees, the cool well water, and the red roses my Mother planted. In the distance horses and cows grazed in fields of alfalfa green. Not far from our home was a Mayhaw Pond and on summer nights you could hear the songs of Whippoorwills in the distance. Often around midnight, I remember the sound of a distant train whistle. Talking about summer reminded me of the hot nights we had back then. Often we would sit on our back porch on clear summer nights and watch shooting stars streak across the sky. We had no electricity in the country and it was before window screens came about, so our home was open to mosquitoes which made the nights miserable.

When we went to church on Sundays I remember the summer heat and the ladies in the pew fanning with straw hand-held fans that advertised products like flour or fertilizer.

There are special memories of the games we played as a child , "drop the handkerchief," "kick the can," and "hide and seek". One time when I was six years old my brother Houston gave me a ride on his new bicycle. He had just received his bicycle as a gift at his nine-year-old birthday party. While we were riding my left foot got entangled in the spokes causing an open wound that re- sulted in a deep infection that lasted all summer. I also remember another problem I experienced as a child, one day while helping my father with our horses a horse whipped his tail hitting my left eye. The tear in my eye was deep and the doctor was concerned that I might become blind in that eye.

As I remember my childhood summers, one of the things our family enjoyed most was making homemade ice cream on Saturdays, especially on "The Fourth of July." One time to surprise my Mother, my Father brought home a wind-up Victrola. We listened to music and watched my parents dance. Sometimes in summer, my older brothers would take the younger ones swimming at a nearby pond. Those were precious childhood memories.

I also remember as a young child working in our family's garden. There was never an end to picking butter beans and peas. As a family, we all worked together in the garden and there were always plenty of tomatoes, sweet potatoes, corn, and okra. We shared our blessings with all of our neighbors.

When I was about ten there was an outbreak of malaria caused by mosquitoes. That summer every afternoon it seemed to rain and with no screens, on the doors or windows our family came down with malaria. Several families in our community died from malaria. One man said the only way his family survived the fever was through prayer.

One fall when I was eleven years old there was a horrible hurricane called The Okeechobee Hurricane that came from Florida. We didn't receive the damage as they did in Florida and on the Georgia coast, but the horrific winds almost destroyed our cotton crop. After the hurricane, I remember everyone helping pick cotton off the damaged stalks; that happened in September of 1928.

In 1929 the economy began to change as the world entered "The Great Depression." Times really became difficult for our nation and as a boy of twelve, I remember my family watching every penny. By the time I was fifteen, people were suffering economically. In the big cities, people were so hungry families turned to soup kitchens just to survive. Being on a farm we always seemed to have plenty of vegetables, milk, and meat. It was a hardship for my Father to buy necessities for our family and farm. I remember my Father saying that many of the banks in our country had closed their doors.

President Herbert Hoover was in office from 1929 to 1933; life in our nation was so hard for families, people nicknamed those days, "Hoover Times."

For me growing up on the farm from age thirteen till age twenty I didn't know anything about politics. My main concern was just helping my family on the farm. In my spare time, there was fishing, hunting, and on Saturdays going to nearby towns like Gordon, or Ashford, Alabama. Of course, on Sunday we attended church and enjoyed Christian fellowship; often having a social gathering where each family brought a special dish for dinner on the grounds.

In 1933 Franklin D. Roosevelt became President, I do remember the excitement of his being elected President, and everyone calling him, "FDR." With President Roosevelt in office people felt there was hope. Our nation was glad when he announced, the " New Deal," which gave the farmers relief and; most importantly the ability to borrow money. He also created "The Civilian Conservation Corps," which gave young single men the ability to earn thirty dollars a month allowing them to send twenty-five dollars to their families, keeping the other five for themselves. The CCC as it was called gave young men the opportunity to learn in a military environment as they worked to improve our nation's forests and parks.

When President Roosevelt spoke on the radio, his talk was called a "Fireside Chat." When he spoke everyone in our family would gather around the radio and listen intently. Talking about the radio, I enjoyed listening to country music; singers like Jimmie Rodgers and Bob Wills were my favorites.

As 1941 arrived I joined the "Civilian Conservation Corps," which looking back, prepared me for my future in the military. Joining the CCCs enabled me to help my family monetarily, it also offered an opportunity for me to learn a variety of new skills, with classes on different subjects. I enjoyed working with new people and the challenge of constructing parks. At our home base, we also had sports activities like baseball, volleyball, and swimming. They also taught wrestling with demonstrations showing us new holds; we even had wrestling matches which I enjoyed. Our CCC camp was near Tallahassee, Florida, and I often went into town for different events, like dances or movies...

The CCC allowed you to reenlist periodically and in the spring of 1941, my family asked me to return home; my Father's health was declining and help was needed on the farm. As I arrived home my Mother cried as she told me how sick my Father was. He was eighty-three years old and his heart was failing. She called it a leaky heart. As I entered my Father's room that day, we cried together as I hugged him. It was painful for me to see him so frail.

I helped with the farm the rest of 1941, and The Lord blessed our family with a great harvest. I also cared for my aging parents, especially my Father. I was truly thankful to be with my family as 1941, ended.

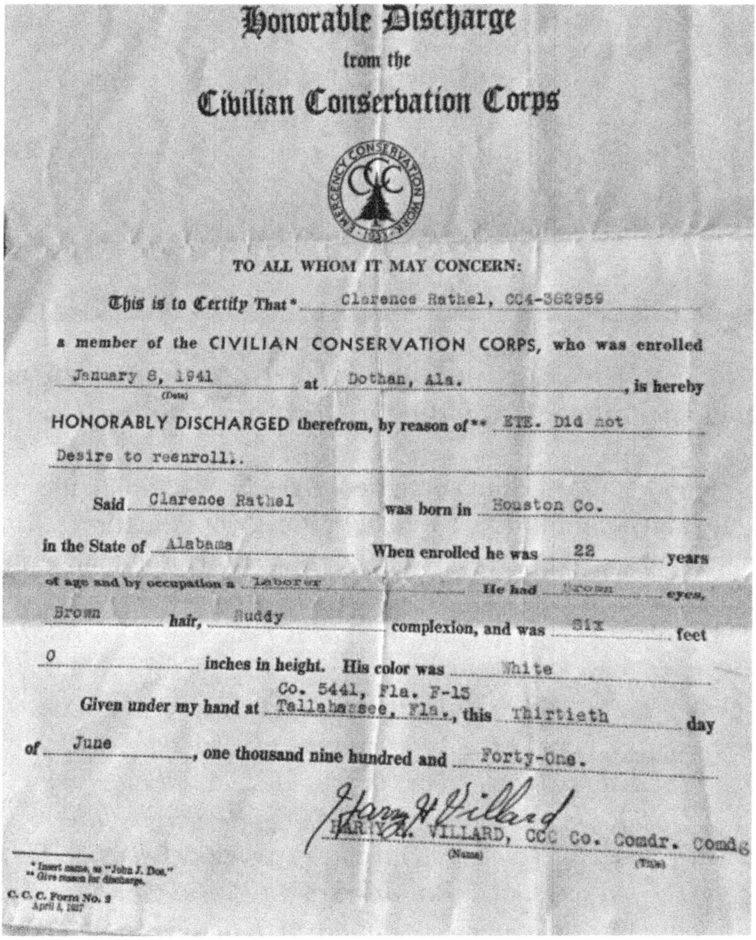

Honorable Discharge
from the
Civilian Conservation Corps

TO ALL WHOM IT MAY CONCERN:

This is to Certify That* Clarence Rathel, CC4-362959

a member of the CIVILIAN CONSERVATION CORPS, who was enrolled January 8, 1941 (Date) at Dothan, Ala., is hereby HONORABLY DISCHARGED therefrom, by reason of** ETE. Did not Desire to reenroll.

Said Clarence Rathel was born in Houston Co. in the State of Alabama When enrolled he was 22 years of age and by occupation a Laborer He had Brown eyes, Brown hair, Ruddy complexion, and was Six feet 0 inches in height. His color was White

Given under my hand at Co. 5441, Fla. F-15 Tallahassee, Fla., this Thirtieth day of June, one thousand nine hundred and Forty-One.

HARRY H. VILLARD, CCC Co. Comdr. Comdg (Name) (Title)

* Insert name, as "John J. Doe."
** Give reason for discharge.

C. C. C. Form No. 2
April 1, 1937

The Honorable Discharge from the "Civilian Conservation Corps" dated June 13, 1941

1941, The Year That Challenged The World.

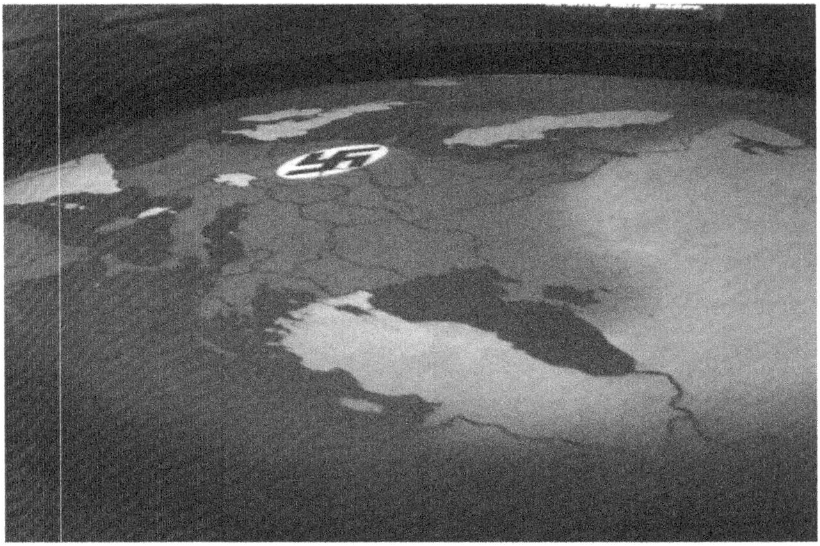

Adolf Hitler's, German Aggression: The World War II Globe, The National Infantry Museum, Fort Benning, Georgia

As my Father's health continued to decline Adolph Hitler was creating havoc in Europe. Ultimately President Franklin D. Roosevelt addressed America and the world about Hitler and German aggression.

"The Only Thing We Have to Fear is Fear Itself" - President Franklin Delano Roosevelt

One of the most memorable speeches President Franklin D. Roosevelt ever gave to the American people and the world was "Fireside Chat number 17, An Unlimited National Emergency," May 27th, 1941. In the speech, he said, about Adolph Hitler's Nazi threats, "The Only Thing We Have to Fear is Fear Itself." This was the second time in his Presidency that President Roosevelt spoke those words; the first was in his first inaugural address on March 4, 1933.

Our family listened with heartfelt concern to every word President Roosevelt spoke on May 27th, 1941. Little did we know that seven months later our nation would enter "World War II."

On December 7th, 1941 the Japanese attacked Pearl Harbor, on December 8th, the United States declared war on Japan, and on December 11th, Germany and the Axis Powers declared war on the United States. Soon after those events, my Father's health became very serious. On December 23, 1941, my Father passed away. His death was heartbreaking for our family, especially my Mother. The grief of my Father's death with the added sorrow of our country being at war made it one of the most disheartening and saddest Christmas times I've ever experienced. The funeral ceremony for my Father was very touching, Pastor Ethridge paid honor to his life and service to our country. After the service at the Gordon Baptist Church, my Father was buried at The Gordon Cemetery.

On certain days the River my Father loved, The Chattahoochee can be seen from his gravesite.

(*The Steamboat Era in Chattahoochee River History began with the 1827-1828 voyage of the "Fanny" from Apalachicola, Florida to Columbus, Georgia. In the 120 years that followed, more than 200 steamboats would ply the rivers waters, transporting people, and connecting the region to world markets, steamboats profoundly influenced the growth and development of the South.)

**The Columbus Museum, Columbus, Georgia*

***

Chapter Two

Midnight Train – Destination 77th Infantry Division United States Army

On September 16^{th,} 1940, our country enacted, "The Selective Training and Service Act." It was the first peacetime draft of men ages 18 to 36, in our nation's history. Men were required to register with their local draft boards and if selected serve one year in the military. When the United States entered World War II the requirements changed to all men ages 18 to the day before their 45th birthday and if selected they were subject to military service. During World War II all men from their eighteenth birthday till the day before their 65th birthday were required to register. Later President Roosevelt extended the time of service to 30 months or longer if needed for national security.

In September 1942 I received notification that I had been selected to defend our country. (My draft notification) My Mother being old wanted me to stay home, (I'm sure all mother's during that time felt the same). She even asked the man in charge of the draft board if there was any possible way he could change things. The man replied, "I wished I could but the answer is no." My Mother's heart was broken so we returned home.

On October 20, 1942, after saying goodbye to my family and hugging my Mother for possibly the last time, I boarded the midnight train from Gordon to Dothan; with other men who had received their draft notice. Upon our arrival in Dothan approximately 50 men from Southeast Alabama, Southwest Georgia, and part of the Florida Panhandle (near Alabama) reported to the officer in charge and were transported in military buses to Fort McClellon, Alabama; for processing and possible selection to serve in the armed forces.

The Greyhound Bus Station...Dothan, Alabama

Once at Ft. McClellon, we were checked physically and instructed to go to a special area when our names were called by the officer in charge. Once those selected were in place the officer announced that we had been tested and were able to serve in the military. He then asked us to raise our right hand and take the Military Oath of Allegiance, (*The Military Oath of Allegiance is very serious and I've always considered it a lifetime commitment*).

At that point, on November 4th, 1942, I officially entered the United States Army. Documents were signed and instructions were given to report back in fourteen days. After fourteen days everyone that was selected boarded a train from Ft. McClellon to Ft. Jackson, South Carolina; for eight weeks of Basic Training. I was assigned to the 77th Infantry Division. After Basic Training during the coldest part of winter, we were transported to Louisiana in January of 1943 for eight weeks of maneuvers.

The "Louisiana Maneuvers" took place in central Louisiana beginning in September 1941 and involved 400,000 soldiers. These maneuvers were the largest training event in U.S. military history. The maneuvers created a mock war in 1941. After the original maneuvers in 1941, the military continued to have maneuvers in Louisiana as it built the Army. The 77th Infantry Division partnered with the 90th Infantry Division in January of 1943 for strenuous maneuvers which simulated battles in the war.

***

Chapter Three

Training in the Sonoran and Mojave Deserts

The next phase of training was for deployment to the deserts of North Africa.

The North African Campaign was a vital part of World War II, at stake for the Allies was control of The Suez Canal and the oil fields of the Middle East. The Campaign was from 10 June 1940-13 May 1943.

More importantly, victory in North Africa would permit a gateway to conquer Sicily and begin the Italian Campaign leading to victory over Germany's main European ally, Fascist Italy.

After the Louisiana Maneuvers, we were transported by train to Hyder, Arizona for desert training in the Sonoran Desert and the Mojave Desert. Hyder was a watering spot in the very center of the Arizona desert. We arrived in March and our training required five-mile marches to begin with and progressed ultimately to twenty-five-mile marches in the desert sand. After several weeks we were transported to Needles, California in the Mojave Desert, (in an area close to Death Valley); for desert training and marches up to twenty-five miles in 120-degree heat in mountainous terrain. We were only given small rations of water and food during these marches. To conquer our thirst we were instructed to place a small stone in our mouth, (this was a survival technique passed down from Native American Indians), it did help, but all I could think about was the cool well water from my home in Alabama. The Mojave Desert is beautiful, I even found sea shells in the sand, but the heat saps every ounce of energy in your body. It was not unusual for men to suffer heat exhaustion on the long marches. We were in the desert's hot blistering sun for seven long months of torment.

The Desert Training Center in Southern California and Western Arizona, was built-in 1942 to prepare several army divisions for deployment to the African Theater. From March 1943 to September 1943 the 77th Infantry Division was based at Camp Hyder. By the middle of 1943, the African War Campaign was well under *control and The 77th Infantry Division's mission was changed to fight in the Pacific*

Theater against the Japanese. (Definition of the word "Theater," when used militarily in a war includes – Air, Sea, and Land that is directly involved in a war. In this case, the War in the Pacific involved one-third of the surface of the world, or 64,000,000 square miles.

Description: Desert Training Center map 1943, US Army
Date: 1943
Source https://apps.dtic.mil/dtic/tr/full text/u2/a537098.pdf
Author US Army
This file is a work of a U.S. Army soldier or employee, taken or made as part of that person's official duties. As a work of the U.S. Federal Government, it is in the public domain in the United States.

The Pacific Theater of War-Twin Drives to Victory

U.S. Pacific strategy enveloped two offensives, one advancing across New Guinea towards the Philippines and another crossing the Central Pacific. The two would converge at the Japanese home islands. On July 2, 1942, the U.S. Joint Chiefs approved a dual offensive plan against Japan. While there would be some early operations in the South Pacific, forces under Admiral Chester Nimitz would attack directly across the Central Pacific from Hawaii. Forces under General Douglas MacArthur would attack across New Guinea and through the Philippines en route to Tokyo. The Chiefs kept to themselves the assignment of tasks and allocation of resources.

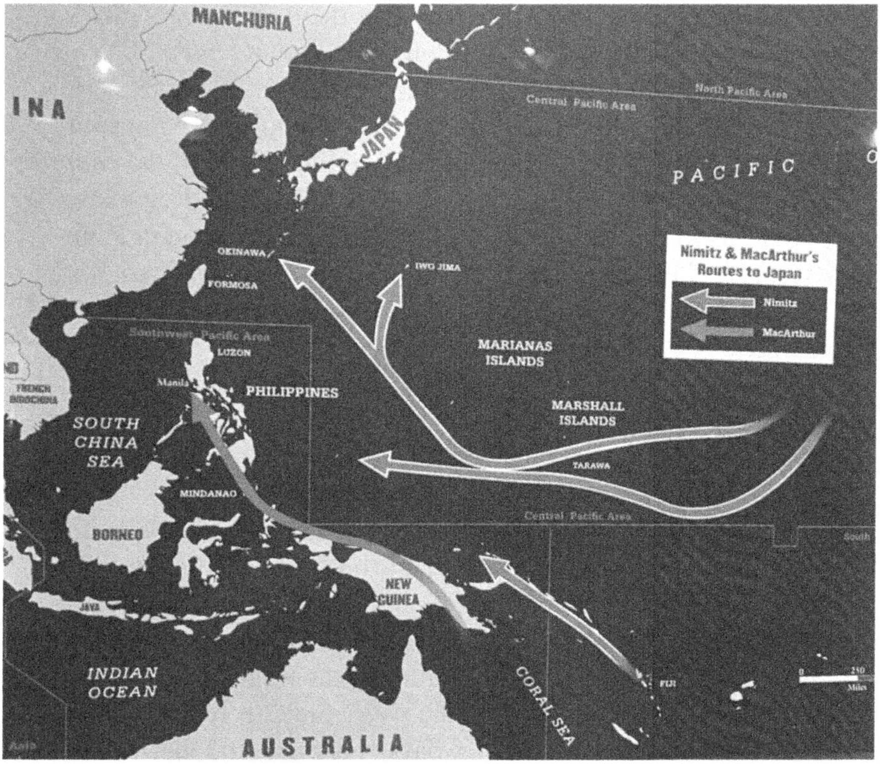

Map of The Pacific Theater of War, World War II, National Infantry Museum, Fort Benning, Georgia

***

Chapter Four

Change of Mission

Near the end of my desert training, there was a mission change for the 77th Infantry Division. On May 13, 1943, the Allies declared a victory in Northern Africa over the Axis Powers *(See note explaining Allies and Axis Powers at end of chapter).* When this happened the war in the Pacific became our Division's priority. New orders were issued and the 77th Infantry Division split three ways for specialized training. *(See note at end of the chapter explaining military organization).* Sending the 305th Regimental Combat Team to Indiantown Gap Military Reservation in Pennsylvania, and the 306th Regimental Combat Team to a West Virginia training area near Elkins. I was assigned to the 307th Regimental Combat Team and sent to Camp Pickett, Virginia, and then Camp Bradford, Virginia to begin amphibious assault training. A schedule of rotation was set for three types of training, mountain training, amphibious assault training, and marksmanship combat training with field exercises.

A priority focus of our training was mastering and using different weapons, especially the M1 Carbine, the M1 Garand rifle, and the Thompson Submachine Gun. Each team rotated every three weeks to the other camps to receive the full curriculum. The number one priority in combat training was weapons. All weapons used by the infantrymen in the war were covered. We were at the Firing Range or in weapons class every day and many nights when in combat training. **This training was very important — in battle, it could save the lives of your buddies, or your platoon!**

Actual cartridges. The one on the left is a .30 carbine as used in the WWII M1 Carbine. The one on the right is an actual WWII .30-06 cartridge as used in the M1 Garand Rifle. The black tip indicates it has an armor-piercing core.

The mountain training near Elkins, West Virginia was very strenuous; it included **mountain climbing and survival techniques.**

An added thought about the intense combat training. It included topographical classes and field exercises using a compass. Knowing how to read a map is one thing, but knowing how to use a map and compass in battle could save lives. Try to imagine one directional mistake that could place you and your platoon in danger. Once again we completed marches from five miles up to twenty-five miles. These were speed marches in mountain terrain in freezing weather. There were also mock battles using all the weapons we had been trained with. These battles included tank warfare and tank destruction, mines, booby traps, grenades, mortars, and machine guns.

Browning Machine Guns, National Infantry Museum, Fort Benning, Georgia

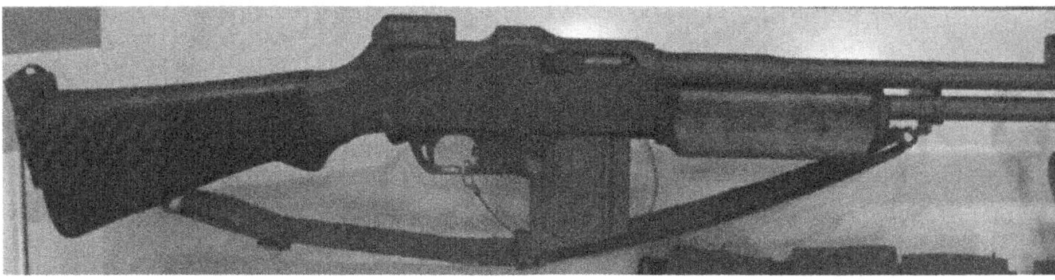

M1918A2 Browning Automatic Rifle (BAR) National Infantry Museum, Fort Benning, Georgia

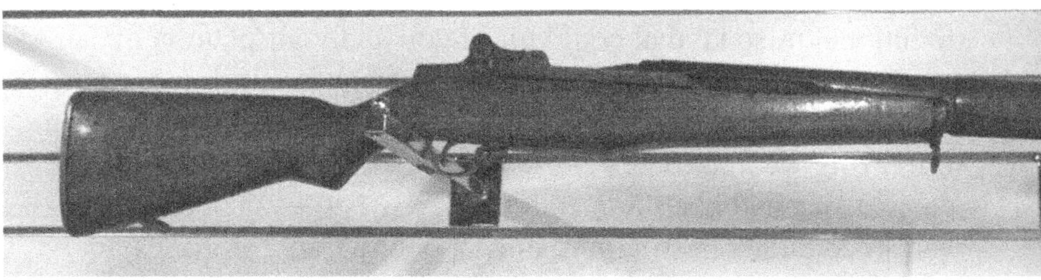

M1 Garand semi-automatic Rifle

"Follow Me," National Infantry Museum, Fort Benning, Georgia (notice the hand grenades draped around the soldiers neck.)

The Dangerous Art of Amphibious Assault:
"A Day That Will Live In Infamy"
On December 8, 1941, President Franklin Delano Roosevelt in his Declaration of War against Japan said these words in his speech to the United States Congress:
Paragraph four of his speech.

"The attack yesterday on the Hawaiian Islands has caused severe damage to American naval and military forces. Very many American lives have been lost. In addition, American ships have been reported torpedoed on the high seas between San Francisco and Honolulu. Yesterday the Japanese Government also launched an attack against Malaya. Last night Japanese forces attacked Hong Kong. Last night Japanese forces attacked Guam. Last night Japanese forces attacked the Philippine Islands. Last night the Japanese attacked Wake Island. This morning the Japanese attacked Midway Island. "

WAR!

SAN FRANCISCO, Dec. 7. President Roosevelt announced this morning that Japanese planes had attacked Manila and Pearl Harbor.

OAHU BOMBED BY JAPANESE PLANES

Pearl Harbor Attack: pictures from, The Museum of Aviation, Robbins Air Force Base, Georgia

When the Department of Defense changed the mission of the 77th Infantry Division to the Pacific Theater, one of the most im- portant skills we were trained for was beach landings under en- emy fire— Amphibious Assault. The Empire of Japan had control of the Pacific and one of the most effective ways to regain control of the islands they had captured was by amphibious assault.

After our Division's Desert Training in Arizona and California, our soldiers were sent to different forts and camps for specialized training with rotation to Amphibious Assault starting at Camp Pickett. Later full training in Amphibious Assault was com- pleted at Camp Bradford, Virginia, and Solomon's Island, Mary- land. Camp Bradford is located near Virginia Beach where the Chesapeake Bay meets the Atlantic Ocean. Solomon's Island is between the Chesapeake Bay and the Atlantic Ocean.

To explain Amphibious Assault our first instructions were about ships, just a few fun examples our instructors gave us in the beginning, "ship terminology;" a floor on a ship is called a "deck,", and a kitchen is called a "galley." There were lectures on ship life, safety, ship courtesy, and what to do when seasick. We slowly progressed to more serious training like landing techniques, climbing down cargo nets suspended from platforms, landing boats, landing plans and diagrams, and combat loading of equipment and troops.

The next phase of training required physi- cal strength with multiple drills climbing down cargo nets, fully loaded with combat gear and weapons. The gear weighed approx- imately 70 pounds. In a real Amphibious Assault, the gear would include a carbine or rifle, a poncho, C rations, tablets to combat malaria, ammunition, several hand grenades, a blanket, mess kit, personal hygiene items, a steel helmet with webbing, extra cloth- ing, entrenching tool, extra boots, and other necessary equipment.

The landing craft used for Amphibious Assault is called a "Landing Craft Personnel Vehicle," (LCPV) or Higgins Boat, each boat carried approximately thirty-six men or a platoon. Soldiers boarded the boat by climbing down cargo nets from a troop transport (a ship). The soldiers exited the boat by charging down a lowered boat ramp. The exit was the hard part, in a real battle, it involved wading in the ocean water at least waist deep to the beach under enemy fire.

The final Amphibious Assault training was a trip up the Chesapeake Bay to Solomon's Island from Camp Bradford and was a mock battle. The beach landing I participated in was in November in horrible weather and freezing ice cold water. Several soldiers in my platoon suffered from hypothermia and those who made it were not allowed to have campfires but endured the night in the woods just as if in true battle. The final training for Amphibious Assault for the 77th Infantry Division before leaving for Hawaii, our next step toward war, took place on the 16th of December 1944.

Due to the intense training of our Division, from the deserts of the Southwest to the icy waters of Chesapeake Bay, some in our Division died from the elements of nature or pure exhaustion. These gave their all, some were my buddies.

I just have to tell you the truth, it was hot, cold, blistering and freezing, thirsty and hungry, mud and ice, everything in between.

Sergeant Rathel was awarded "The Combat Infantryman Badge," before his first battle
in World War II. A Star was added for The Korean War.

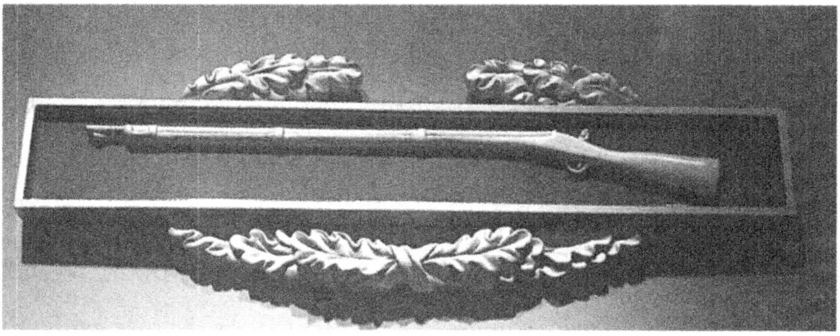

The Combat Infantryman's Badge is one of the most treasured military decorations a soldier can be awarded. The design of the Combat Infantryman's Badge symbolizes the qualities, sacrifice, experience, and history of the many foot Soldiers that have faced death for the United States, The Oak Wreath symbolizes steadfastness, strength, and loyalty. The Model 1795 Springfield musket represents the first official U.S. Infantry weapon and the 3-inch long Infantry blue rectangle superimposed with the rifle signifies the uniqueness of the infantry-the U.S. Army branch that is the- " Queen of Battle."

The Allied Powers and Axis Powers WW II:

The Allies were an international military coalition formed in 1941 by the United Kingdom, the United States, the Soviet Union and China. Those were the principle countries. Other countries were part of the Allies called Allied with governments-in exile: These countries were Poland, Czechoslovak, Norway, the Netherlands, Belgium, Luxembourg, Free France, Ethiopia, Greece, Yugoslavia, the Philippines, India, Canada, Australia, New Zealand, South Africa, Brazil, Mongolia, and Mexico. As the war evolved some of the conquered countries became allies.

Top left to right – President Franklin Delano Roosevelt, The United States of America. Sir Winston Churchill, Prime Minister of The United Kingdom. Joseph Stalin, General Secretary of The Communist Party of the Soviet Union. These were the principal leaders of the Allied Coalition during World War II.

The Axis Powers were a military coalition between Nazi Germany, the Kingdom of Italy, and the Empire of Japan. These were the three major countries. Other Axis states were the Kingdom of Hungary, the Kingdom of Romania, the Slovak Republic, the Kingdom of Bulgaria, the Independent State of Croatia, the Republic of Finland, and the Kingdom of Thailand.

·om left to right- Adolf Hitler Chairman of the German Reich, Benito ʼussolini, Prime Minister of Italy, Emperor of Japan, Hirohito.

ıese were the principal leaders of the Axis coalition during World ʼar II.

ʼictures of World War II leaders- The National Infantry Museum, ›rt Benning, Georgia

nformation about Allies and Axis powers from different historical urces and museums

This is an example of how an Army Division was organized during 1944-1945 and is an overview only

(please note the Army is very flexible and can shift soldiers, equipment , and heavy weaponry very quickly based on the situation)

A division had between 14, 250 to 15,000 soldiers which consisted of officers and enlisted men.

There were three squads each composed of riflemen and one light ma - chine gun, commanded by a sergeant.

.

The three squads made up a platoon of approximately forty men and were commanded by a second lieutenant.

Three rifle platoons plus a weapons platoon (machine guns) made up a company, approximately one hundred ninety men commanded by a captain

.

Three rifle companies plus a weapons company (motors and machine guns) with a headquarters company made a battalion, about eight hundred seventy men, commanded by a lieutenant colonel.

Three battalions together with artillery and supporting units made a regiment, approximately three thousand one hundred twenty men, (3,120), commanded by a colonel.

The complete division consisted of three regiments plus artillery and engineer, signal, and medical, commanded by a major general. Once again a division had approximately 14,250 -15,000 soldiers.

End of Example (Example taken from different articles on the military organization)

There were ninety-one total Army Divisions in World War II. The Army had six Corps and twenty-one Divisions in the Pacific Theater plus six Marine Divisions. Corps is two or more Divisions combined to form Army Groups.

Please note that the Air Force was part of the U.S. Army during World War II. The United States Air Force of today was separated from the Army on September 18, 1947.

Special notes: The United States Coast Guard and Merchant Marine' s played a vital part for the Allies during World War II, transporting troops,

supplies, and equipment around the world in dangerous war zones, many lost their lives under attack. Several ships were also lost at sea and some crews were captured and taken as prisoners of war.

The following information was obtained from "The National World War II, Museum," New Orleans, La.

There was 8,267,958 total Army personnel, 474,680 total Marine personnel, 85,783 total Coast Guard personnel and, 380,817 Navy personnel for a Total American Armed Force of 12,209,238 at the peak of World War II, 1945.

This picture of Sergeant Rathel as a Private was taken before deployment to The Pacific Campaign in 1944.

These are the Insignia worn by Sergeant Rathel during World War II

The 77th Infantry Shoulder Sleeve Insignia

Special Unit Designation "Statue of Liberty"

Motto, "Ours to hold it High"

The 307th Infantry Regimental Insignia

Motto, "Clear the Way"

***

Chapter Five

Hawaii-Jungle Warfare Training 1944

As February arrived the 77th Infantry Division assembled at Camp Pickett, Virginia, and began to prepare for overseas travel. Our orders were to be kept secret. We were allowed to send our last non-censored letters home along with cameras and radios, which were not allowed on our mission. Physical checkups including dental were given to everyone along with required shots. All uniforms and clothing items were brought up to standard. Equipment and weapons were inspected and made ready for travel.

Our Division traveled by train to San Francisco, between March 8-13 and we reassembled at Camp Stoneman, California. Once everyone arrived we were transported to the docks in San Francisco harbor and boarded several grey Navy transports (ships) which remained docked overnight.

The next day as we sailed out of San Francisco under the beautiful Golden Gate Bridge, this thought pulled at my heart, for many even myself this could be our last farewell. It took several transport ships to carry our complete division, and we were escorted by several heavily armed Navy ships. Once we were in the Pacific Ocean the water became turbulent. Orders were given for blackouts at night, the only light visible was the moon and stars, something I would become used to in war.

After a few days the beautiful coastline of Oahu,Hawaii could be seen in the distance, then Honolulu and Diamond Head. We finally docked in Oahu and eventually, my regiment the 307th camped in the sugarcane fields at Kahuku, in the northeast corner of Oahu. After our camp was set up we were able to relax for a couple of days and many of us went to Waikiki Beach. We were like tourists enjoying the beach swimming and eating Hawaiian food.

In Oahu, our training began at the Unit Jungle Training Center. There was hand-to-hand combat training with instructions in martial arts like Karate, and Jujitsu. Survival training where we searched for food, set booby traps, and camouflaged ourselves.

Sneak attack techniques, guerrilla warfare, and instruction on what to do if abandoned or lost in the jungle. One important thing we memorized was what to do if captured by the enemy as a prisoner of war.

We practiced how to survive a ship disaster or explosion at sea by jumping from a high platform into a lake of fire, then swimming under the oil and gas as it blazed above. We continued to do amphibious assault training using an (LVT) Landing Vehicle Tracked (a troop carrier designed for the coral coast of many of the islands in the Pacific). If you can imagine jungle warfare and everything involved from mine detection, and mine placement to understanding how snipers hide and kill, we did it.

Occasionally as a break from training the USO would have entertainment. They would bring in different entertainers or country music groups, which was really appreciated by everyone.

Now a few words about Jungle Warfare and the other challenges we faced in the jungle; Insects, diseases, and Wounds. We were vaccinated for tetanus, typhoid, smallpox, cholera, and yellow fever before we left Camp Pickett, Virginia. In the jungle, it rains daily causing a soldier to be constantly wet. A soldier's feet if not kept dry are subject to Jungle Rot, a debilitating fungal infection. In the jungle clean drinkable water is a must, we were given tablets to put in our canteens to purify our water. If your water was contaminated you were subject to dysentery. Another issue in the jungle is food poisoning caused by contaminated food. Mosquitoes are prolific in the jungle, not only do they sting and irritate they carry Malaria and other diseases. Something very important to every soldier is wound care. Proper wound care is essential in battle, for yourself and others.

***

Chapter Six

The Emotions of Battle and War

The following is added by the author from personal conversations with his Father and others who have experienced the battlefields of War...from a Christian perspective.

***Fear** is a natural emotion for a soldier before and in battle. The relationship a soldier has with God and the strength of their faith is tested with every move when under fire. Prayers asking God for protection and guidance before, during, and after a battle are made. Many soldiers make promises to God while under the stress of battle, some keep their promises to live a righteous life, and some try to forget their promises when the battle is over. Some soldiers gain more trust in God through the experience of battle and unfortunately, a few lose their faith in the battle. Some soldiers have empathy with their enemy, realizing their opponent is much like them, in a war and a place not of their choice. A certain conflict exists in a soldier's mind about killing another person in battle, which presents a mental and spiritual dilemma for some. Soldiers of Faith pray before the battle and know God is with them. Many in their prayer ask God for Grace and Mercy while making peace with God. In the end, War is one of the greatest tests of faith God presents to mankind.*

***Anxiety** like fear is something every soldier experiences before a battle and a person of faith deals with it differently than one who doesn't have faith. Soldiers of Faith confront anxiety with scripture- "The Word of God," many quote verses in their prayers before and during the battle. Scripture has a way of giving hope, encouragement, and purpose to a soldier regardless of rank, or circumstance.*

***Depression** is often experienced by soldiers after the battle, possibly caused by the horrible sights of death and destruction they have witnessed; or possibly grief caused by the loss of a friend or members of their squad or platoon. Loss is the big reason for depression. Wounds, injury, and physical distress caused* *by being in an unfamiliar place in unsanitary and unhealthy circumstances can cause a soldier's depression. Sometimes making a quick decision in a panic situation is relived in a soldier's mind or his dreams. In the end, depression has many causes, and even though a soldier returns home after the war, sometimes he carries in his soul the memories of battle and his feelings of helplessness for life. Today the term for this emotional stress is called, PTSD (Post-traumatic Stress Syndrome).*

***

Chapter Seven

Prelude to The Liberation of Guam

Sergeant Rathel, Speaking

The 77th Infantry Division was on combat-ready alert to invade Guam in late June 1944. The scheduled time of the invasion of Guam hinged on the Battle of Saipan, an island approximately 140 miles north of Guam. The Battle of Saipan was primarily to cut off the route and flow of enemy Japanese soldiers to Guam. The battle was to be over in three days but lasted approximately fifteen days. So intense was the Battle of Saipan that our partner division, the 27th Infantry Division, along with the 2nd Marine Division and the 4th Marine Division experienced extremely high casualties. The Battle of Saipan ended on July 9, 1944. This was disappointing news to hear as we prepared for the Battle of Guam.

(In the Battle of Saipan more Americans were killed and wounded than in any other battle in the Pacific Theater of War during World War II.)

Map of Guam detailing the chronological order of The Battle of Guam World War II,

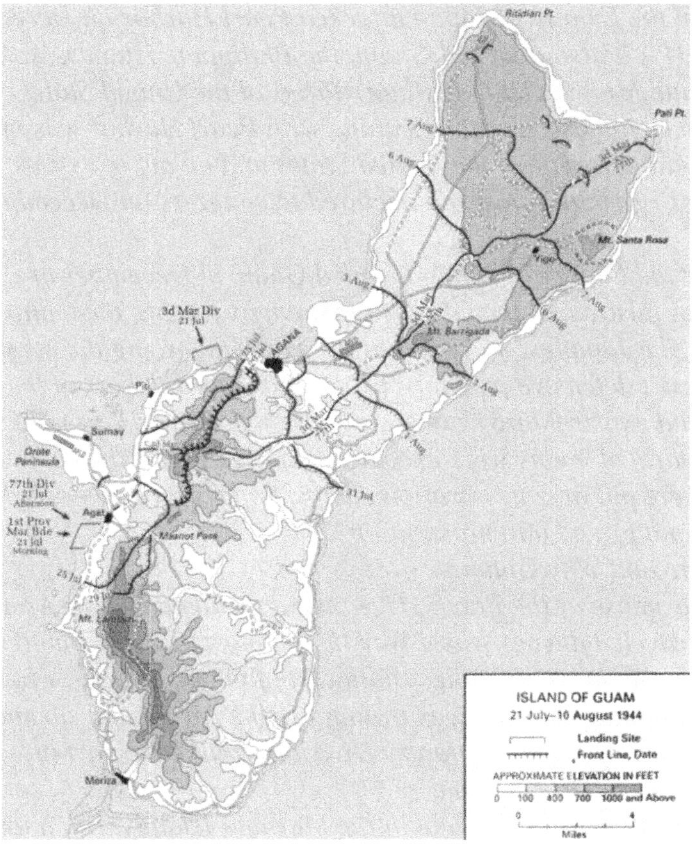

July 21, 1944–August 10, 1944

Description: Map of the Battle of Guam, 1944 During the Pacific Campaign

This image is a work of a U.S. Military or Department of Defense employee taken or made as part of that person's official duties. As a work of the U.S. federal government, the image is in the public domain, the United States.

If you remember I spoke of my Father Honeywell Rathel, fighting in the Spanish American War, which ended with "The Treaty of Paris," signed on December 10, 1898. In the treaty, Spain also ceded possession of Guam and Puerto Rico to the United States and allowed the United States to buy for Twenty Million Dollars, the Philippine Islands. When that happened Guam and the Philippine Islands became United States Territories.

When the Empire of Japan attacked Pearl Harbor on December 7, 1941, it also attacked Guam, the Philippine Islands, Wake Island, and Midway Island, all territories of the United States. The attack on these Territories along with Pearl Harbor was officially an attack against the United States and an act of war as President Franklin Roosevelt declared to congress on December 8, 1941.

After the Empire of Japan invaded Guam in December of 1941 they mistreated the citizens of Guam by forcing them into slavery. The Japanese were brutal taskmasters forcing the people to construct defensive positions by digging cave bunkers in the jungle and constructing runways, roads, and airfields. Families were separated, many were executed, some were imprisoned, and others were put in concentration camps. Some women were held captive and forced into prostitution.

Where and Why Guam

The expanse of the Pacific Ocean and the distance our Armed Forces traveled during World War II, is almost beyond comprehension. Guam is 3,400 miles southwest of Oahu, Hawaii, where the 77th Infantry Division was training In the spring and summer of 1944. In July the 77th Infantry Division would take part in—the Battle to Liberate Guam.

Guam is the largest island in the Mariana Islands with a land area of approximately 215 square miles or about the size of half of the Florida Panhandle. Guam's coastline of seventy-eight miles is surrounded by coral reefs, It has steep coastal cliffs with narrow coastal plains in the north surrounded by thick jungle; in the south are mountains with Mt. Lamlam having a height of over 1300 feet. From a military observation, the island's terrain allows very limited access for Amphibious Assault. The island was formed by volcanos and is the largest island between New Guinea and Japan. Its value to the Empire of Japan was fresh water and a supply base for Japanese ships. On Guam's Orote Peninsula the Japanese had an airfield that was used to rearm and refuel their carrier-based aircraft in attacks against the American Fleet. Strategically the Japanese knew Guam was important and could be used by the United States as an air and naval base to launch future attacks against Japan using the new B-29 Superfortress

Bomber.

B-29 Superfortress Bomber, Range 5,592 miles, Top Speed 365. Picture from the Museum of Aviation, Robbins Air Force Base, Georgia

Knowing the value of Guam to the United States the Japanese began to prepare well in advance for an invasion. They had excellent observation points from mountainous lookouts and knew that the landing topography for Amphibious Assault was very limited, placing obstacles like mines, explosives, and barbed wire in the coral reefs at potential landing sites, which even at low tide our soldiers would have to wade in chest high water to access the beaches. The total number of Japanese Army was estimated to be close to 22,000 on the island in June of 1944. They planned to form a web, set up a crossfire, and wait until our infantry made it to the beach, then obliterate our soldiers with a storm of untold firepower. The Japanese had the perfect defensive position for battle.

The following are the United States ground forces who Liberated Guam in the Battle of Guam

July 21,1944-August 10, 1944.

III Amphibious Corps, 3rd Marine Division, 77th Infantry Division, 1st Provisional Marine Brigade,

**Total Force approximately 59,000*

** Japanese Total Force approximately 19,000*

** Figures from multiple historical articles and history books* Exact numbers are unknown*

On July 18th the advance Division Headquarters and the 305th Regimental Combat Team of the 77th Infantry Division with the III Marine Amphibious Corps sailed from a holding area in Eniwetok Lagoon (a very small island about 1000 miles from Guam) arriving near Guam before daylight on the 21st of July.

I was with the 307th RCT (Regimental Combat Team) along with the remainder of the 77th Infantry, in a convoy of several transport ships from the island of Oahu, Hawaii. We paused at Eniwetok and loaded LVTs (Landing Vehicle Tracked) for our assault. After quickly loading the landing craft we sailed. The plan was for us to arrive at high tide about noon behind the first invasion at 0830. En route, we reviewed our final assault plans going over maps and models of the island, and rechecked our weapons. (While saying a prayer)

The delay caused by the battle of Saipan, (referred to in a previous paragraph) allowed our Navy to attack Guam longer and with heavier bombardment from battleships and destroyers than would have been possible otherwise. The bombardment which started in the first few weeks of July was coordinated with intense air strikes from the Army Air Force to weaken the resistance of the enemy in advance of our invasion.

***

Chapter Eight

The Battle to Liberate Guam

On the morning of July 21st, 1944 at 0530* the Battle to Liberate Guam Began

At 0530 along the western coast of Guam and inland, the thunderous sound of the United States Navy's big guns started raining firepower equivalent to a storm of destruction the mind cannot phantom. The heavy bombardment continued until 0800 when the bombardment intensified but changed directly to areas along the shoreline, at the same time waves of aircraft rained down rockets along the beaches and hills nearby. The first Marines were in their landing craft ready for the assault, as they approached the beaches 1000 yards away rockets from the Navy saturated the coast trying to clear the way for the assault. The attack was synchronized and the Navy shifted the bombardment above the coastline trying to shield as many soldiers as possible from enemy fire once the Marines were within 300 yards of the beaches.

**Military uses a 24-hour time system: example midnight is 0000 hours 11 pm is 2300 hour*

This is a very brief overview of the Amphibious Assault and Objective

Geographically Guam is an oblong island about thirty miles long, having a peninsula named the Orote Peninsula which is just below the center point on its western side. The peninsula was inaccessible for Amphibious Assault due to very high cliffs on each of its sides. The Japanese had an airfield on the peninsula which was well fortified. To the Northside of the peninsula stood Apra Harbor and to the Southside was Agat Bay. The Amphibious Assault was carefully planned for one force of the III Marine Amphibious Corps (the 3rd Marine Division) to approach from the north on the beaches of Asan just north of Apra Harbor, and the other force of the Marines (the 1st Provisional Marine Brigade) from the south on the beaches of Agat Bay near the village of

Agat. The two Amphibious Assaults were to take place simultaneously. Once on shore the Marines would link together at the rear of the Orote Peninsula and secure that area from the rest of the island, this would be accomplished as the 305th RCT of the 77th Infantry gave support. Altogether there would be multiple waves of Amphibious Assault that would continue as the ground was secured from the enemy, and as relief was given to the first warriors. The 307th RCT which I was part of was held in reserve, along with the 306th RCT and the remainder of the 77th Infantry until activated as the battle expanded in the days ahead.

Due to the excellent bombardment of the Navy and air strikes by the Army Air Force, many of the Japanese attack positions on the coast had been nullified, which helped our Marines and soldiers as they beached. Unfortunately, they still experienced machine guns, explosive land mines, and mortar fire from pockets of the enemy, along with heavy weapon fire from the hills. Overall our forces made it to the beaches and began to take ground very slowly under fire. The end objective (which would take several days) of the assault was for both forces from the north and south to meet at the rear of the Orote Peninsula, secure the area and once the Armies were in place begin an offensive attack of the peninsula. This move would require the rear of the offensive to be defended by bringing in the 306th RCT and the 307th RCT along with great coordination from all forces, including Marines, Navy, Army, Army Air Force, and Coast Guard. After the Orote Peninsula was secured the rest of the island would be taken one sector at a time. An added challenge in the mix of all this was the harsh environment of the jungle, rugged cliffs, hills, and eventually the mountains. Please remember this was just an overview and in the end, it came with a heavy cost, the cost of many lives, I'll talk more about that later.

Example of an Amphibious Assault:
Picture: "Tip of the Avalanche," The 36th Infantry Division lands at Salerno, Italy on September 9th, 1943

National Infantry Museum, Fort Benning, Georgia

***

Chapter Nine

My First Amphibious Assault in Battle

If I take wings of the morning, and dwell in the uttermost parts of the sea;
Even there shall thy hand lead me, and thy right hand shall hold me.
Psalm 139: 9-10 (The Bible NIV)

At this point, I will discuss my personal experience starting from the day and time I left the transport ship. Please remember I was in Company A of the 307th Regimental Combat Team, 77th Infantry Division, and I was there.

July 24th, 3 pm (1500 military time)

Amphibious Assault, White Beach 2 the coast of Agat Bay

I along with the rest of the 307th RCT disembarked a transport ship during a turbulent storm at sea. While transferring from the ship to the landing craft (both in motion) and on the nets, two soldiers slipped falling into the ocean, being lost forever in its waves. There were many landing craft filled with soldiers, and because of the swells and turbulence, we were unable to immediately approach the coral reefs. Several men became seasick. Once to the reefs, the landing crafts were unable to proceed because of obstacles requiring everyone to unload in neck-high water with waves. In the water our footing was unstable due to the jagged bottom of the coral reef, occasionally someone would step in a hole and go under, requiring others to help. Each soldier was weighted by at least 80 pounds of gear. Some men were carrying extra equipment like parts of machine guns or mortars, most of us carried sacks of extra hand grenades which were hung around our necks, along with our weapons and a grenade launcher, of course, we tried to keep our weapons dry. We were dropped about 800 yards from the beach and under the constant threat of enemy fire as we arrived.

After wading and swimming to the shore in the evening mist, we found the beach littered with hundreds of dead Japanese and American soldiers. The Navy was still pounding the enemy with heavy fire above the coastline, and the Army Air Force swarmed above with air strikes less than a mile away. We slowly moved inland about 300 yards from the shore and found an old Marine dugout position. The Marines had moved further ahead in the battle. We were their replacements, they had been in battle since day one. (approximately four days)

Our first night was miserable, we were wet, discouraged, and unable to sleep because of enemy fire. We had nothing to eat except rock candy. During the night word came that the Marines from the north invasion had connected with the Marines of the south invasion at the base of the Orote Peninsula.

As daylight arrived and the morning beach fog lifted we beheld the most horrific sight the mind can imagine, hundreds of bodies strewn along the coastline covered with flies; with the heat of the day the smell was humanly unbearable and sickening. Eventually, the United States Naval Construction Battalion (nicknamed the Seabees) would use bulldozers to bury the bodies, of course, there was a procedure and burial would be accomplished with professional dignity.

Later in the morning, the order came for us to push forward, we were to replace a Marine Regiment on the frontlines. The operating procedure in battle, whether on a small scale or large scale, is to always have a reserve force available as a replacement for the frontline force, this was a very large scale. The reserve always stayed in the rear yet had to follow along. In this war, the enemy also attacked from the rear. We were the reserve for the Marines and as we replaced them they would become our reserve. Ultimately my regiment the 307th RCT would play a major role in securing the base of the Orote Peninsula, by holding the link of the north and south original Amphibious Assault Armies (Marine and Army) at the rear of the peninsula. This was a vital area that came with the cost of many lives.

As we moved forward in the jungle we encountered invisible snipers, hidden in the trees, or perfectly camouflaged underbrush, their bullets came unexpectedly and killed with lightning speed. We encountered heavy resistance in several places, booby traps, mines, and machine gun nests, always moving forward. As we advanced we encountered several dead soldiers. These were our men from the 77th Infantry, as the battle permitted they eventually would be lined in rows near a path. Some of the bodies were swollen and well I won't say anything else. (I saw this several times in this war) I felt sorry for the men assigned to the burial registration team; they had the duty of taking the dog tags (identification tags) from around the neck of the dead soldiers and preparing their bodies for pickup. They would first cover the body with creosote (a tar-like substance) to preserve it, then after the substance hardened take a poncho (a waterproof covering) and wrap the body. They would then place one dog tag on a small wooden cross at the head and the other dog tag on a toe. Eventually, the human remains would be loaded onto a truck and at some point buried in a temporary military cemetery in Guam. The dog tags would be used to mark the grave and to notify the soldiers loved ones.

July 26th Attack of the Orote Peninsula

At daybreak, on the morning of July 26th, the Arterially of the 77th Infantry began the battle for the Orote Peninsula. Several 155mm Howitzers, the big guns, were fired, along with many tanks; the Navy and Marines joined the attack with their big weapons. Above the Army Air Force swarmed dropping bombs in waves. This created an inferno of unbelievable magnitude. For the Japanese Orote was their most fortified fortress. That night the battle would become extremely intense as the 1st Provincial Marine Brigade with other Marine forces began to move forward from the base of the peninsula. The Japanese had the rear well defended, with several well-constructed pill boxes, (a pill box is fortified and contains machine guns fired through openings), trenches, booby traps, and other elaborate defenses. As the Marines conquered this area moving forward the Japanese realized they were trapped. Their next move was a suicide attack which resulted in all the enemies defending the base of the peninsula being killed by the Marines. The battle for the airfield on the Orote Peninsula would continue for the next three days with heavy resistance.

During this time I was in the jungle at the base of the Orote Peninsula near Mount Tenjo (a mountain over 1000') with part of the 307th RCT. We were on high alert, our duties were to secure and patrol a wide area with two objectives stop bands of Japanese intruders from the main island (which reconnaissance showed to be in the thousands). Watch for the enemy coming from the peninsula our way (which some did). It had been over three days since we had eaten real food. You can only imagine the weakened state we were in. Finally, we were able to stop for a few minutes and eat K-rations which tasted like paste, they had a hard biscuit and some type of rubbery meat, no taste, and of course a piece of rock candy.

K rations were supposed to hold you till you had B or C rations which were just a little better. Fortunately, at some point a pallet of B rations arrived, I think they were supposed to go to another Regiment but unfortunately, they were not around so our men, the men of Company A 307th Regiment requestioned them, in other words, we took them. The meals were just navy beans and some type of meat, but we didn't care we were starving in this battle.

July 28th Mount Tenjo—From Out of the Blue—Friendly Fire

On July 28th Company A of the 305th RCT was sent on a very important reconnaissance mission. Their purpose was to clear Mount Tenjo of enemy resistance. While on the way to the top, several snipers were encountered and killed. Once arriving at the summit, suddenly out of the blue, the patrol was attacked by friendly fire from the Army Air Force. Fortunately one of the privates risks his life by running to an open area and displaying the platoon's signal panels. The pilots seeing the panels on their next attack immediately stopped firing and veered off. Thankfully no one was injured. In the end, Mount Tenjo was declared clear of the enemy which was the final link in securing all the immediate area behind the Orote Peninsula. Later that evening the 2nd Battalion of the 307th RCT relieved Company A of the 305th RCT and secured all of Mount Tenjo. In the end, everyone agreed that the friendly fire experience would never be forgotten.

On the same day near the base of Mount Tenjo, our regiment set up a small bivouac area to alternate patrols and secure the area. Around the bivouac area, we placed guards. Occasionally one of our guards would encounter a sniper, or there would be an enemy intruder. That afternoon one of our guards seeing suspicious movement near the edge of our area, "accidentally", shot and killed a red cow thinking it was the enemy. That evening we made a fire and everyone had a piece of steak (beef). About the time we finished eating the Colonel arrived and joined us. I think in my whole life that was the best meal I've ever had. I was famished!

On my father's tape about this incident he was very specific about the red cow – for those of you who love Biblical symbolism, the Red Heifer has great meaning. I'm almost certain he was unaware of this significant symbolism, at this time in his life. Randy Rathel

July 29th— Reconnaissance

On July 29th the Battle for the Orote Peninsula was over and the airfield was declared free of enemy resistance. As that battle was drawing to a close and knowing that only a small portion of the enemy would be conquered on the peninsula, important decisions had to be made by our Generals about enemy locations and movement on the main island of Guam. Due to the massive foliage of the jungle It was impossible to observe the enemy from the air. The only way to discover their movement and location was by reconnaissance. Five reconnaissance teams of approximately four men each accompanied by a native guide were sent in different directions from the headquarters area to gather information and radio back their observations.

***

Chapter Ten

The Torch of Liberty—Liberation of a Concentration Camp

July 30th

Reconnaissance revealed the enemy was shifting northward into a deeper jungle, possibly where supplies and armament were hidden. It was decided by Major General Bruce 77th Infantry Division and Major General Geiger of the Marine Corps to reposition the Army and Corps in preparation for a northern attack. The attack would be led from the southeast by the 77th Infantry Division with the 1st Provisional Marine Brigade attached in Corps reserve and from the southwest by the 3rd Marine Division.

Before the attack could happen the 77th Infantry would move across the island to the east, clearing the area of resistance, then advance north clearing resistance. Once arriving in the north the 77th Infantry Division and 3rd Marine Division would attack the enemy together. The first part of this repositioning move would be accomplished on foot, hiking and marching through very difficult terrain by the infantry accompanied by light tanks and minimal motorized vehicles. The second part would be accomplished by the Seabees quickly building roads with bulldozers and checking bridges while under guard. The third part of the move would be to bring in heavy equipment, and trucks with supplies including ammunition, food, and medical needs. This was the basic plan and the unknown factors were enemy resistance, weather, and difficult terrain.

July 31st

On the morning of July 31st, the 307th RCT and the 3rd Battalion of the 305th RCT led the advance of the 77th Infantry in a march that began from an area near Mount Tenjo behind the Orote Peninsula just below the center of Guam on the western side to the Pago River, on the eastern side. This march of ten miles was very difficult taking us across terrain filled with cliffs, gorges, jungle, hills, and swamps. Each man was loaded with heavy gear, about seventy pounds. Besides the terrain, we constantly battled swarms of mosquitoes, heat, and jungle rain.

As we marched patrols on our perimeter and watched for enemy resistance, none was encountered, allowing us to make excellent time. At midday, our command, who had planned for a two-day march realized we were close enough to the Pago River to continue.

Approaching the Pago River, Company L of the 307th came into contact with a few Chamorros (native citizens of Guam) who told our soldiers of a nearby concentration camp called Camp Asinan. With caution our patrol checked for enemy guards finding they had left, leaving about two thousand people abandoned and incarcerated. Our soldiers immediately set these people free. The joy on the faces of these people made the men of the 307th RCT smile. Many of the captives especially the men had been tortured and everyone suffered from malnutrition. The children were very frail and those with families were able to reunite. Our soldiers quickly shared what few food items they had in their pockets and many of those liberated said they had prayed for a long time that the Americans would come. These people were very thankful and offered to help our men, but it wasn't necessary. Their gratitude was overwhelming. *Some even had tiny U.S. Flags they waved, hidden from their captors for years. The experience of liberating these people gave our soldiers a renewed purpose and defined Freedom in a living breathing way we all understood.

The shoulder sleeve Insignia of the 77th Infantry Division is, "The Statue of Liberty," holding the Torch which Lights the Way to Freedom. The motto of the 77th Infantry Division is, "Ours to Hold it High."

Paragraph 3 page 67, "GUAM, Operations of the 77th Division," Center Of Military History, United States Army, Washington D.C. 1990

M5A1 Stuart Light Tank
Crew of 4
Speed 36 mph
Main Armament 37mm M6 in Mount
Secondary Armament three .30 Browning M1919A4 Machine Guns

M4A3 Sherman Medium Tank Crew 5
Speed 22-30 mph
Main Armament 75mm gun M3 (Cannon)
Secondary Armament one .50 caliber Browning M2HB Machine Gun
Two .30 caliber Browning M1919A4 Machine Guns
**National Infantry Museum, Fort Benning, Georgia*

Chapter Eleven

The Crimson Battlefield of Barrigada

After losing the battle on the Orote Peninsula the Japanese left a very small force in the southern half of Guam while retreating northward. In their retreat, their losses continued to accelerate. They knew they would be outnumbered at Barrigada so they prepared a Crossfire Trap……

August 1, 1944

I will always remember this next event because of the thirst we as soldiers experienced, and the many lives we lost in the battle for the water of Barrigada.

When you are thirsty you think of many things, one of my thoughts was of the cool well water on my family's farm in Gordon, Alabama. Oh, how I had wished for just one swallow from that well as we marched north from Pago Bay toward the village of Barrigada. Often thoughts of the beautiful Chattahoochee River that my family loved would appear in my mind's eye. Then I would return to reality while filling my canteen from a dirty stream and adding purification tablets as the sounds of military movement echoed in my ears.

About seven miles from Pago Bay where the 77th Infantry Division, was on August 1, 1944, is Mount Barrigada, which is about 650' above sea level. At its base is the village of Barrigada which had a well and pumping station that provided the only fresh drinkable water for miles. As the 77th Infantry Division and Marines moved northward in our pursuit of the enemy, water became a priority. There were no streams in northern Guam; where the Japanese would eventually make their last stand.

Late in the afternoon of August 1st, the 307th RCT captured the road south of the village of Barrigada. Reconnaissance soon revealed the enemy had control of the village. During the night our commanders planned the attack for the next morning.

August 2, 1944

Early in the morning on August 2, 1944, a column of light tanks traveled the Agana Road toward the village of Barrigada.

The tanks were on a reconnaissance mission checking for enemy positions when suddenly machine gun fire erupted from all directions. The tanks immediately fired back with a barrage of machine gun fire but were unable to completely knock out all the suspected positions. The tanks soon returned to report and await further orders.

The order to return and engage the enemy was quickly given. On their return, the column of tanks encountered heavy resistance, as a large group of Japanese tried to approach one of the tanks with hand grenades, while others fired machine guns from the dense jungle. The tank crews fought back showering the area with machine gun fire and quickly killing those with hand grenades. What was strange is the enemy was so well camouflaged they were almost impossible to locate. Our tanks returned to rearm and start again. This was just the beginning of a hard-fought battle.

A line of attack was formed by the 305th RCT, 307th RCT, and Company D of the 706th Tank Battalion. (in reserve and on the far left was the 3rd Marine Division on a different mission). Our approach from the south, east, and west was in the form of a semicircle at least two and a half miles wide (with a few gaps due to terrain) and five hundred yards away from the center of the village, which had an open field two miles wide bordered by jungle as you entered the village. Light tanks helped give cover as each battalion inched forward in the jungle and on the road.

As each battalion on foot approached the village, mostly in a low crouching position, trying to find cover, we encountered heavy sniper fire. Soldiers fell as if hit by lightning. The snipers were so well camouflaged they were invisible, some even fired from the trees behind us. Suddenly crossfire from machine guns erupted. Then enemy tanks hidden in the buildings of the village began to attack. At one point everyone became disoriented as men fell wounded or dead, everything seemed to swirl in a blur of confusion. We returned fire with our light tanks and advanced slowly with our machine guns ablaze. Later in the afternoon as we gained ground my company, company A with companies L and K of the 307th RCT became pinned down in the very center of the battle. Eventually, our medium tanks which had been delayed en route, arrived giving everyone relief. The battle continued into the night. The next morning reconnaissance discovered the enemy had retreated during the early morning hours completely leaving the village. The enemy in their rush to escape left the well and pump undamaged.

In the end, the water of Barrigada was costly, the price was a battlefield covered in crimson.

The Battle of Barrigada, August 2, 1944,*

**On August 2, 1944, the 77th Infantry Division had 29 killed and 98 wounded.*

**Breakdown -The 305th RCT had 7 killed and 35 wounded. The 307th RCT had 22 killed and 63 wounded.*

Paragraph 1, page 100, "GUAM, Operations of the 77th Division," Center Of Military History, United States Army, Washington D.C. 1990

***

Chapter Twelve

Dengue Fever

July is the beginning of the tropical rainy season in Guam. After our battle for water in the village of Barrigada which ended on August 3rd, we moved into the very dense jungle of Mount Barrigada. The jungle was so thick we had to cut our way through with machetes. With daily heavy downpours of rain and jungle humidity, we were constantly wet. Our clothing and weapons became filthy with the red volcanic mud that the island is made of. The real plague we faced was mosquitoes especially in the evenings and at night. It wasn't fun being in a wet sloppy foxhole, fighting mosquitoes and wondering if the enemy would suddenly attack.

Our command estimated the Japanese enemy remaining on the island at this point to be less than five thousand.

We were constantly moving forward trying to keep the enemy off balance and on the run. As the 307th RCT conquered Mount Barrigada the enemy continually attacked with snipers, and ambushes. Mount Barrigada is covered with dense jungle and our command had hoped that in taking it we would be able to observe enemy movements; but as it turned out that was impossible, the jungle was just too thick.

On August 6th, I was a short distance from our Chief of Staff, Colonel Douglas McNair when he was shot and killed in an ambush attack by enemy soldiers.

Soon after that incident and as Mount Barrigada was under control I became sick with Dengue Fever.

You may remember me talking about mosquitos in the jungle, one of the diseases they carry is Dengue Fever. I came down with it, as well as our whole company. If you can imagine a very high fever, nausea, vomiting, joint pain, headache, and rash I had it. One of the things it causes is tremendous pain behind your eyes, that was my main problem, everyone was deathly sick. Some in our company had other issues like yellow jaundice. It takes about ten days to recover from Dengue Fever. I'm almost certain some of our soldiers died during the outbreak it was so severe.

In my fever, I thought I heard the voice of Tokyo Rose. She was telling me the Americans were going to lose the battle. She constantly talked on the radio when we were en route to Guam on the transport ships, so I guess her propaganda subconsciously worked while I was deathly sick. Everyone knew she was lying. Anyway, Dengue Fever is something horrible. As I healed I found out the Battle of Guam had ended. Tokyo Rose was wrong America had successfully Liberated the citizens of Guam.

*This Rose is not for Tokyo Rose but for the citizens of Guam who suffered persecution and death; the approximately *five hundred Prisoners of War the Imperial Japanese sent to Kobe, Japan after conquering Guam in January 1942; and the *American Armed Forces our country sacrificed while liberating Guam.*

**approximately 1780 American Armed Forces were killed with an estimated 6,000 wounded, average numbers from different historical accounts.*

***

Chapter Thirteen

Prelude to the Battle of Leyte

The Bataan Death March- April 1942

In April 1942, American Soldiers and Philippine Citizens became prisoners of Imperial Japan and were forced to march from the southern end of the Bataan Peninsula to Camp O'Donnell, a prison camp on the island of Luzon, resulting in an estimated 10,000 deaths en route.

Many POWs endure the worst part of their captivity during transport from the battlefield to prison camp.

Few suffered more than the defenders of the Bataan peninsula in the Philippines during the Second World War. Approximately 10,000 American soldiers, along with 62,000 Filipinos, were forced to surrender to Japanese authority on April 9, 1942. Having already survived months of rough fighting on minimal rations, they were scarcely prepared for the ordeal that lay ahead. How many men died on the Death March will never accurately be known. Rough estimates put American deaths at 600 or 700; Filipino deaths, somewhere between 5,000 and 10,000.

**Prisoner of War Statement, and Photograph of The Bataan Death March: National Prisoner of War Museum, Andersonville, Georgia*

October 20th, 1944

United States General Douglas MacArthur was the Supreme Commander of Allied Forces Southwest Pacific Area during World War II. General MacArthur, felt it was a moral obligation to liberate the Philippines from the oppression of the Imperial Japanese and returned to Leyte accompanied by four 6th Army Divisions. This action alarmed the Imperial Japanese which resulted in The Battle of Leyte Gulf.

Picture from The National Infantry Museum, Fort Benning, Georgia

Having vowed to return after his evacuation from the Philippines in 1942, General Douglas MacArthur

made good on his promise when he waded ashore at Leyte Island on October 20, 1944.

October 23-26,1944

The Battle of Leyte Gulf

The Largest Naval Battle in History according to many historians — The Battle of Leyte Gulf, October 23 to October 26, 1944, was the first battle the Imperial Japanese used kamikaze (suicide) attacks with aircraft. The Imperial Japanese realized that if they lost Leyte they would lose the war.

***

Chapter Fourteen

The Land Battle of Leyte

"Follow Me! " By H. Charles McBarron, Jr. On October 20 1944 Colonel Aubrey A. Newman, commander of the 34th Infantry Regiment leads his men in establishing the first foothold in the Philippines. As the fighting rages on Leyte Beach, he orders his men to " Get up and get moving! Follow Me!" The U.S. Army had returned to retake the Islands."

Picture from The National Infantry Museum, Fort Benning, Georgia.

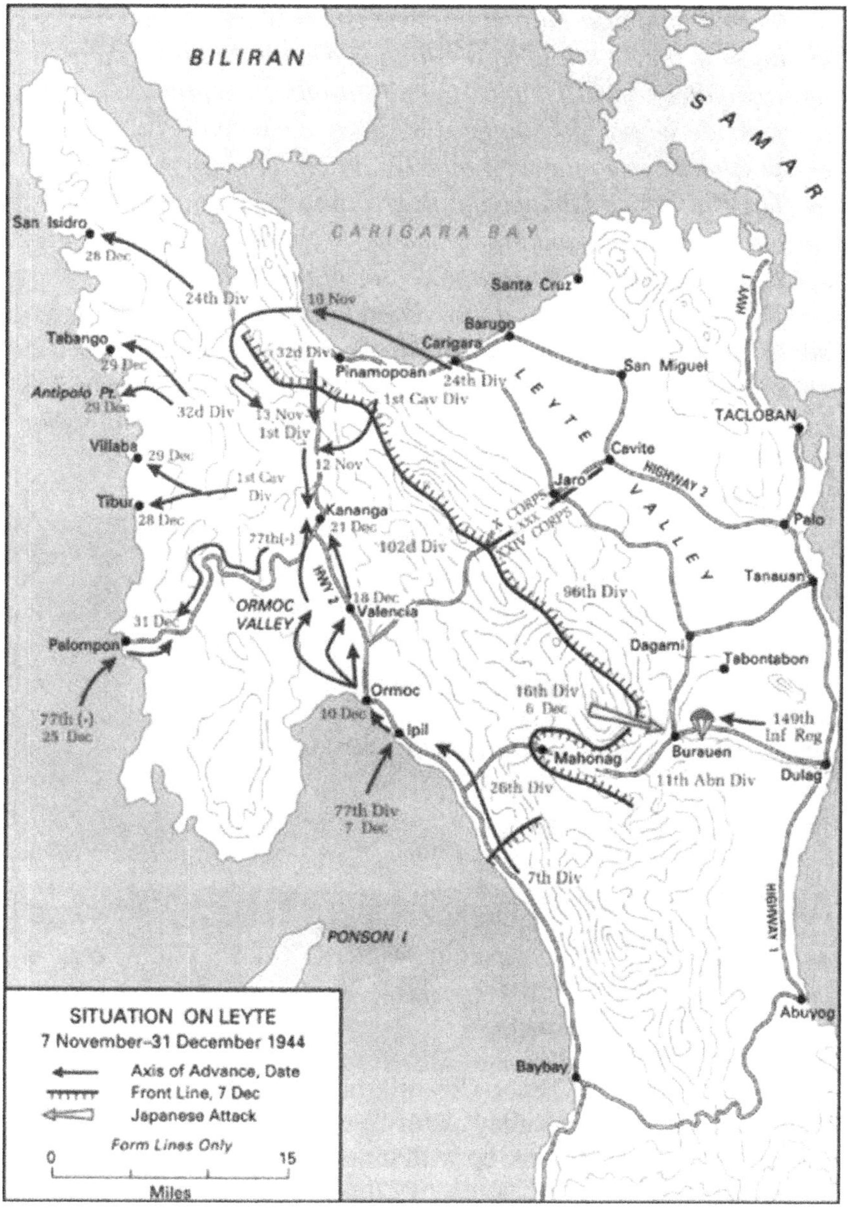

Description: Battle of Leyte, Situation 7 November to 31 December 1944
Source:http://www.ibiblio.org/hyperwar/USA/USA-C-Leyte/maps/USA-C-Leyte-5.jpg
Author: no author information

This image is a work of a U.S. military or Department of Defense employee, taken or made as part of that person's official duties. As a work of the U.S. federal government, the image is in the public domain in the United States.

Geographic information of Leyte Island:

Leyte is the eighth largest Island in The Philippines, there are approximately 7,641 islands in the Philippine Archipelago, which are considered part of Southeast Asia, it is located between The South China Sea to the west and The Pacific Ocean to the east.

The Philippine Islands are divided into three island groups: Luzon, the Visayas, and Mindanao.

The island of Leyte is in the center of the Visayas.

The island of Leyte is approximately 110 miles long north to south and about forty miles wide at the widest point. Leyte is mountainous in its interior with a valley in its Northeast which is an agricultural area. The island is heavily forested, has monsoon seasons and is subject to periodic typhoons.

The hand-painted American flag was made by POW Curtis G. Davis in Camp #10 D in Tokyo. In violation of camp authority, the American POWs stood at attention before the flag and recited the Pledge of Allegiance on New Year's Day, 1945. In Curtis' words, "I was a very, very proud American Army private that day I can assure you. Some time later the Japanese found the flag, confiscated it and threw it in the office trash. One of the prisoners, emptying the trash found it and returned it to me."

Flag and Statement from Curtis G. Davis, American Prisoner of War in Japan during World War II National Prisoner of War Museum, Andersonville, Georgia

Sergeant Rathel, speaking……

On November 3rd, 1944 the 77th Infantry Division quietly sailed from Guam. Filled with patriotic emotion, I watched as the Flag of the United States of America faded into the horizon. Would this be my last vision of Guam? Our destination was New Caledonia, an island in the Coral Sea. After Guam, our division needed a time of rest. That was the plan - rest. New Caledonia is three thousand miles from Guam, and as we sailed we listened to Tokyo Rose. She played American music, often interrupting her broadcast with propaganda, saying, the United States was going to lose the war. I guess no one told her about the estimated eight- een thousand Imperial Japanese killed in Guam or the fact that the island had been liberated.

Sometimes things don't work out the way you plan. Just a few days away from New Caledonia, on November 11th, our com- mand received new orders, we were needed four thousand miles away in Leyte, an island of the Philippines.

We did manage to stop briefly on the island of Manus. We ar- rived two days before my birthday on November 17th, 1944. While there I pondered my twenty-seventh birthday; who would have believed it would be on a small island in the Pacific Ocean near Papua New Guinea. Then the thought occurred, I was born during a time of war, World War I. My birthday reminded me of my family back in Alabama. Reminiscing I thought of my dear older sister Estelle, who passed away in 1930. Estelle had married a gentleman named Sherman Butler, Sherman had been a warrior in World War 1 in the famous 42nd Rainbow Division. Come to think of it General Douglas MacArthur was the one who had the idea to form the Rainbow Division, and now I was under his leadership. He was the Supreme Commander of Allied Forces Southwest Pacific Area. As I thought of Estelle and her sudden death I remembered I was fourteen when she died; leaving behind her only child, a five-year-old son named Harris. I loved Harris and he became my little brother. When Estelle passed away Sherman was torn with grief; can you imagine coming from the battlefields of death in World War 1, finding the love of your life, and then suddenly she's gone? I especially remember my Mother's tears, our family's sorrow, and the question we all asked, Why? Estelle's death came in the middle of the Great Depression and Sherman had to leave in search of work, so my Father and Mother adopted Harris, and our family loved him.

On November 19th, my birthday, we sailed for the east coast of Leyte in a large convoy of transport ships. Our destination was an Imperial Japanese-controlled territory. Knowing our battles would be intense our chaplain frequently had prayer services on deck. Everyone prayed together, everyone..............

On November 23rd, early in the morning, the 77th Infantry Division arrived in Leyte. The 307th RCT was directed to disembark in Dulag a port town on the east coast of Leyte Gulf. Once there our orders were to move inland to LaPaz, a town southwest of Dulag. The east coast of Leyte had been captured by the 6th Army and the 77th Infantry Division wasn't expected to enter combat immediately but that's not what happened...

They say first impressions are lasting impressions; my first impression of Dulag and Leyte was the mud. We arrived during monsoon season, with torrential rains every day. On November 30th the 307th RCT was temporarily split with orders to patrol and guard different naval airfields. The challenge for our commanders was the movement of supplies and heavy artillery. The roads and bridges were practically impassable due to the torrential rain and mud.

After enduring patrols in the rain with occasional skirmishes against the Japanese, our division started reassembling on December 1st, on the beaches of Tarragona and Rizal. It was a massive effort to position weapons, supplies, and men for a planned Amphibious Assault on the west side of Leyte, our objective was Ormoc City. (Ormoc City, had a very important port that was a pivotal point the Imperial Japanese used to supply their forces, naval, air, and, army, it was well protected.)

On December 6th, approximately eighty vessels accompanied by Navy destroyers began the 225-mile move around the southern tip of Leyte, and up the west coast. Our convoy of vessels arrived at Deposito, a few miles south of Ormoc City at sunrise on the *seventh of December.

At exactly 7:07 am, Navy destroyers began to bombard the coast, softening enemy positions, before our assault. While still in the ships the Japanese Air Force arrived; just as quickly our Army Air force arrived flying P-40 Warhawks in opposition. Overhead there was a dogfight (air battle). On the ships, we prepared for the assault.

(*Pearl Harbor was attacked on December 7th, 1941)

The Curtis P-40N Warhawk was equipped with six M2 Browning .50 caliber machine guns, three on each wing, it also carried bombs. The Museum of Aviation, Robbins Air Force Base, Georgia.

The Amphibious Assault of Deposito and The Village of Ipil

Our amphibious assault of thousands of soldiers was backed by heavy naval bombardment above the shoreline. As we approached the beach in knee-high water the 307th RCT (my regiment) advanced slowly under enemy fire. Moving inland we encountered intense enemy resistance. Our tanks and Naval bombardment allowed our advance, but suddenly the Japanese air force strafed (attacked with bombs and machine gun fire) along the shoreline.

As we overcame the attack my regiment slowly moved forward toward Ipil, a village less than a mile from the beach. Then suddenly enemy machine gun nests hidden under native huts erupted with a storm of fire. In the intense battle, we returned fire killing over eighty enemy combatants. After the battle, we secured Ipil by destroying the enemy's supplies, ammunition, and equipment.

(The following is for those who desire to read the Amphibious Assault battle plan for this battle more in-depth. For others you

may skip to the next paragraph – Randy Rathel)

For clarity of the battle plan, imagine a semicircle, the center point of the diameter is the point of Amphibious Assault, from that point the different regiments expand; in this battle, part of the 307th RCT advanced horizontal from the center point east, the other part northeast; the 305 RCT expanded east-southeast and the 306 RCT expanded south touching the beach. The depth of the semicircle should always be between 750 yards to 1000 yards. Placed in certain positions relative to the advance is heavy artillery support; in this Amphibious Assault, there was Naval and Army Air Force support. Once an area is secured in this case Ipil, supplies, ammo, and a mobile hospital are positioned. After preparations are made for a possible counterattack reconnaissance teams are deployed beyond the edge of the semicircle. With reconnaissance information in hand, a plan of attack is formed. From Ipil the next objective was Camp Downes en route to Ormoc City the main objective. As the 77th Infantry Division moved along the west coast northward to Ormoc City, advancing from the east coast was the 7th Infantry Division which encountered heavy enemy resistance; delaying their arrival for the planned attack on Ormoc City. The decision for Major General Bruce was (A) to wait on the 7th Infantry, or (B) to proceed with the advance while the 77th Infantry had the element of surprise. One important thing that Major General Bruce had to consider before moving forward was protecting the area already won from a counterattack. To accomplish that part of the force must remain behind.)

The decision was made to stay in Ipil for the night and watch for an enemy counter-attack. Our regiment controlled a large area of the beachhead, and preparations were made for the next day's battle.

We were told our antiaircraft guns had shot down five enemy aircraft. We also learned a convoy of Japanese ships bringing reinforcements to Ormoc City had been sunk by our Army Air Force; unfortunately, four of our transport ships had been sunk by an enemy aircraft attack as they returned to the east side of the island.

During the night Major General Andrew Bruce decided to direct the <u>307th RCT forward,</u> attacking the enemy the next day;

surprising them before they had time to recoup from our invasion. Backing the 307th would be the tremendous heavy firepower of the 77th Infantry's amphibious tanks and several 155mm Howitzers. We would be headed north toward Camp Downes a position controlled by the enemy. The terrain was muddy rice patties, lined with trees sporadic hills, and ridges; we left behind a small force to protect the beach we had won.

An M114 155 mm Howitzer in firing position

Date Sept. 1985 Source DOD Media No Author given

This file is the work of a U.S. Army soldier or employee, taken as part of that person's official duties. As a work of the U.S. Federal Government, it is in the public domain in the United States.

The Battle Approaching Camp Downes

Early in the morning, of December 8th, we moved forward toward Camp Downes. Overcast skies, clothes wet with sweat and humidity, rice patty mud, the smell of artillery smoke, with the thunder of artillery fire, our soldier's bodies took on the smell of War. All around me during the day were men suffering from battle fatigue and heat exhaustion; the battle before we reached Camp Downes was fierce, lasting till sundown. Our antiaircraft weapons were able to shoot down three enemy planes as they strafed the battlefield. Yes, soldiers were killed, I don't remember how many. At the end of the day, we were thankful for the three doctors and medics who worked tirelessly and saved at least four hundred severely wounded soldiers. Our adversary was relentless and reminded me of a bulldog that wouldn't let go.

The Battle of Bloody Hill

On December 9th, early in the morning, the 307th became involved in another all-day battle, we called it the "Battle of Bloody Hill." Camp Downes was on a ridge that had previously been an outpost for the Philippine army before the Imperial Japanese took Leyte. During the night of December 8th, the Japanese sent in reinforcements. Once again our adversary had the advantage, there was no cover for our men to hide behind, it was an open field in front of a tall ridge. Added to the confusion of this battle were rain and mud. It was a machine gun, rifle, and mortar battle, As the day progressed we finally gained ground foot by foot, and we took the ridge of Camp Downes. Yes, it was bloody and once again many were wounded. As we regrouped we could see our objective, the City of Ormoc in the distance.

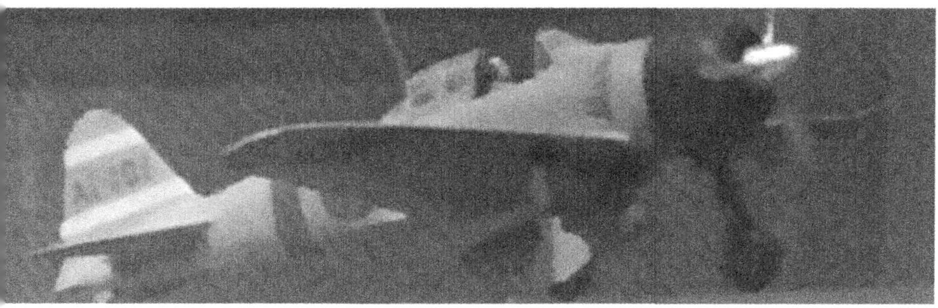

Imperial Japanese Mitsubishi A6M2 "ZERO" Carrier-Based Fighter

The Museum of Aviation, Robbins Air Force Base, Georgia

The Battle of Ormoc City

December 10th, at daybreak our army was strafed by a wave of Japanese Zeros (attack aircraft) which included bombers. Our antiaircraft weapons fought back with a barrage of firepower shooting down one plane which flamed out of control over my position. I watched as it crashed into the ocean. Unfortunately, there were casualties from the bombing. Our command at that point ordered the attack on Ormoc City. We attacked with several

155 Howitzers, Long Tom's, amphibious tanks, and all the firepower the 77th Infantry possessed. Ormoc City became an inferno as the enemy's ammo dump, and fuel storage tanks exploded. Enemy resistance from fortified positions was intense. Our Army Air force strafed Ormoc City relentlessly, and the Navy blasted the port. The Japanese resistance was so intense and well fortified that as the day progressed the 307th battled house to house sometimes shooting from the hip. Yes, there were casualties and many seriously wounded but we gained an offensive position and took Ormoc City by nightfall.

The Church and The Cross

On the morning of December 11th, our soldiers found the City of Ormoc obliterated, except for one building, a church. The windows were blown out but the church was intact. The steeple of the church had a cross on top. As our soldiers secured what was left of the City everyone in my team paused in front of the church. While looking at the cross I felt the presence of God and said a Prayer.

Weeks later our Chaplin mentioned the church of Ormoc City and the cross on its steeple in a Christmas service for our soldiers before we left Leyte. In his sermon he spoke of God's amazing Grace and how God's Son Jesus died on the cross because of God's great love for us and how on the cross Jesus paid the price for our sin reconciling each believer individually to Him. Then the Chaplin said, "if we repented and believed with faith in Christ as our Savior we would have eternal life." Many soldiers trusted Christ that day as their Lord and Savior.

On the night of, December 11th, the action was in Ormoc Bay and its harbor; three vessels bringing Japanese reinforcements and tanks were spotted by our men. Then at just the right time our big Howitzers began firing, two vessels exploded in a ball of bright red fire, and the other vessel was sunk. Later after interrogation, the surviving prisoners informed us they didn't realize Ormoc had been captured by the Americans. *(from late October to December 11th the Imperial Japanese had brought approximately 34,000 soldiers through the port of Ormoc to fight against the Allies).*

Valencia to Palompon

Our next objective was seven miles northwest of Ormoc City to the village of Valencia. Our purpose was to push the Imperial Japanese from our area on the west side of Leyte northward while other divisions pushed them north from the east. Our goal was to ultimately take the complete island from the Imperial Japanese.

On the way, to Valencia, the terrain was uphill mountains with ridges, down through rivers, rice patties, and rain forests. We had to fight for every inch of ground. Those seven miles involved the hardest battles I've ever experienced. One enemy tactic we encountered was snipers in "spider holes,"(a spider hole is when an enemy soldier hides in a camouflaged hole in the ground and fires at you undetected). Another tactic the enemy used was to attack our tanks with "satchel charges," (dynamite or C4 plastic explosive in a bag). The satchel would be thrown into the tank tracks and then explode, causing the tracks to roll off, or if the explosive was more powerful it would destroy the tank.

Banzai

On December 20th, the 307th RCT battled over **2000 of Imperial Japan's elite soldiers;** their 1st Division and the 5th Regiment of their 8th Division in an area north of Valencia. During this battle, two counterattacks of about 200 Japanese in "Banzai" style battle charges took place. (Banzai means, "Long live his Majesty the Emperor," and is screamed by the enemy in a wild attack). By the end of the day over 1400 Japanese were killed while we advanced a mere thousand yards. This battle continued into the night.

A Christmas Victory

Our final battle was on Christmas day when we conquered Palompon the last remaining port on Leyte. The next day the 26th of December we had Christmas Dinner on the beach of Ormoc Bay after our Chaplin spoke of the true meaning of Christmas.

Sometimes in my dreams to this day-I awaken startled-hearing, a Japanese soldier scream "Banzai."

As I ponder memories of Leyte I must say many hardships were experienced. Of course in each battle lives were lost and many were wounded. One of my fond memories of Leyte is the kindness of its people. An example of their kindness was a group of Philippine children who offered their Water Buffaloes to help us across a river with our equipment. This happened more than once.

**The Fight for Leyte began on the 20th of October, 1944, and by Christmas Day enemy losses exceeded 56,000 killed and injured. American losses 3,000 killed and missing. The Japanese were devastated in the Philippines but they continued to fight.*

**Figures from The National Infantry Museum*

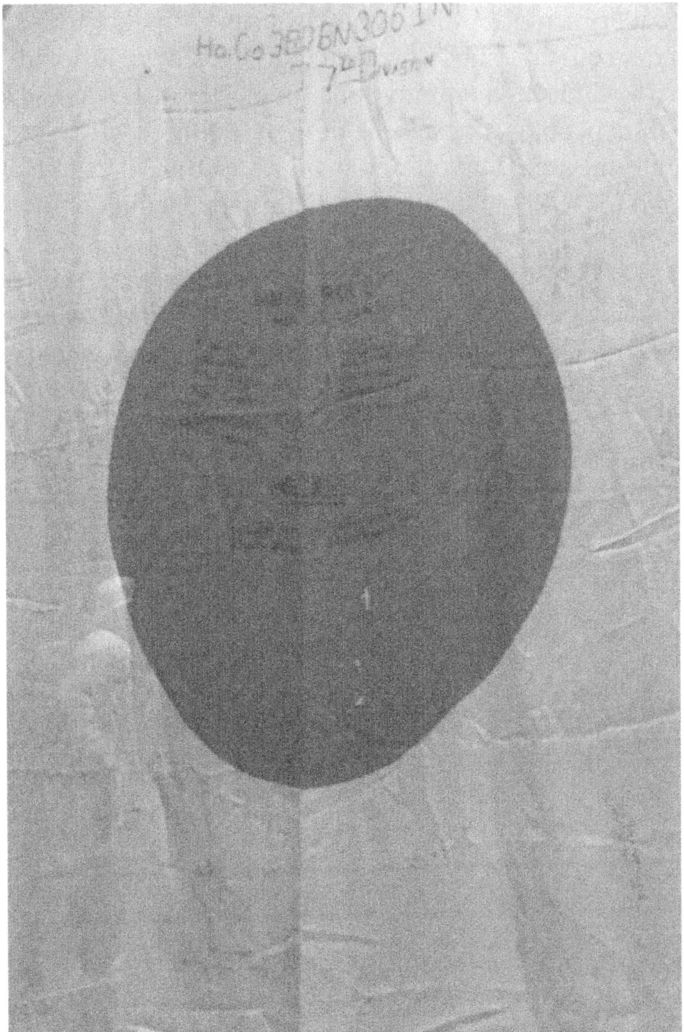

Hinomaru – National Flag of Japan

William Herren, Headquarters Company, 3rd Battalion, 306th Infantry Regiment, 77th Infantry Division obtained this flag in the Pacific Theatre. He honored Soldiers of his company killed in action by writing their names in the red circle. Soldiers from his company autographed the white field of the flag.

**National Infantry Museum, Fort Benning, Georgia*

***

Chapter Fifteen

Prelude to The Battle of Okinawa

It would be impossible for me to describe the panoramic scenes discovered while studying the Battle of Okinawa. The battle's importance to World History and Western Civilization is phenomenal. With great emotion, I now share with you one of the major battles in World History from my Father's eyewitness account, The Battle of Okinawa.

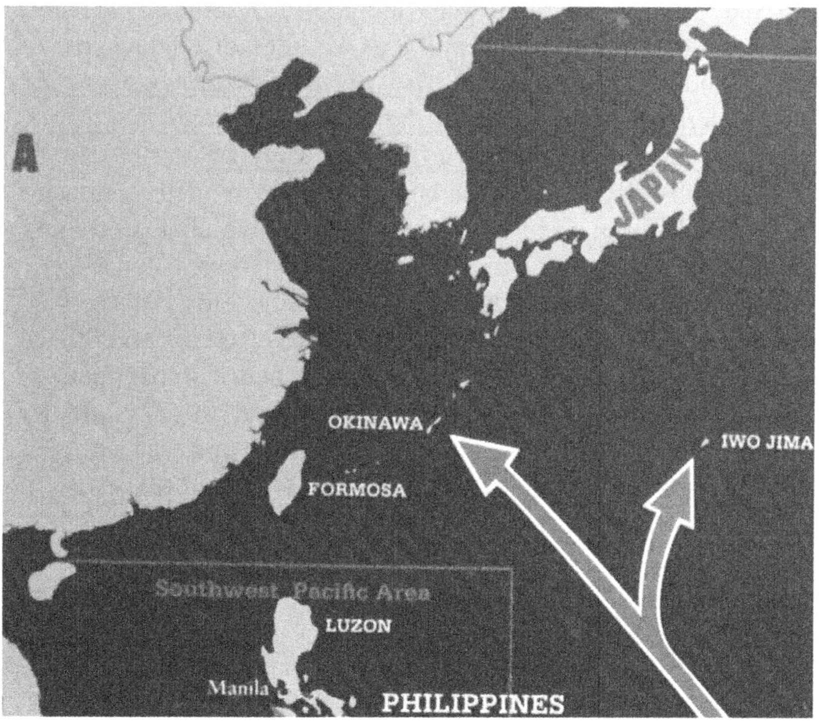

*Map showing Okinawa-*National Infantry Museum, Fort Benning, Georgia*

Sergeant Rathel, speaking

March 1944, The Kerama Islands – First Land Attack on Okinawa

We continued to encounter small pockets of Japanese resistance on Leyte as 1944 ended and 1945 began. Near the end of

February 1945, our next mission was announced, Okinawa, Japan.

In the middle of March, the 77th Infantry Division sailed for the Ryukyus Islands, a small group of islands about twenty miles west of Okinawa, the largest island in the group. Our first objective was a smaller group of islands in the Ryukyus Islands, called the Kerama Islands. The capture of these islands would be our first step into Japan's inner circle of defense and provide our Navy with an operational base for the battle of Okinawa, which was scheduled to begin on April 1st, 1945.

Our arrival at the Kerama Islands on March 26, completely surprised the Japanese, who were unprepared defensively. That morning five battalions of amphibious assault teams attacked five of the islands simultaneously. ***(This was the first attack on Japanese Homeland soil and is considered the first attack by U. S. forces in the Battle of Okinawa in World War II.)***

Geographically the Kerama Islands are surrounded by treacherous ragged Coral, with small sandy beaches which ascend into very rugged mountainous terrain with ridges. The resistance my team encountered was machine gun fire from enemy pillboxes hidden in cliffs. As our platoon advanced we also encountered Banzai attacks, as the enemy wildly approached with bayonets and machine guns, screaming Banzai. In the end, the 77th Infantry Division conquered each island, and what we found was unreal. (The first Japanese mass suicides in World War II. (*Due to the sensitive nature of this topic, the author chooses not to provide details.)*

We also discovered that the islands were used as hideouts for Japanese suicide boats. (A suicide boat was a single-person boat, propelled by a small gasoline motor, used to attack a ship with explosives.)

As we conquered each of the small islands, we searched every crevasse, cove, and cave finding approximately three hundred suicide boats. We also discovered through recovered documents that the majority of the enemy had been transferred to Okinawa for its defense, anticipating the arrival of American ground forces.

From Out of The Clouds- The Kamikaze

April 2, 1945

Sometimes I awaken during the night and in my mind's eye relive this most tragic event. Deep within my soul and spirit I still see the explosions hear the thunderous crashes and smell the smoke, of the Kamikaze.

While the 77th Infantry Division was in a flotilla awaiting its next move, Japanese Kamikaze pilots, and enemy submarine attacks were a constant threat. We were always on alert!

On the evening of April 2nd, while on guard duty on my transport ship, which was part of a large convoy sailing near the Kerama Islands; suddenly from out of the evening clouds we were attacked by a wave of eight Kamikaze pilots. As I helped sound the alarm on my transport, our antiaircraft gunners began to fire, hitting one of the planes directly over the Flag Ship of the 307th Regiment. In flames it exploded as it crashed into the deck of the ship. During the attacks four more of our transports were crash-dived… as our ships fired our Air Force engaged and the Kamikaze left. During the night we were notified that some of the best leaders of the 307th had been killed, these were men I knew and respected, yes there was grief and anger- because of the Kamikaze.

Picture: The Museum of Aviation, Robbins Air Force Base, Georgia

The Battle of Bloody Ridge - Ie Shima

In the East China Sea four miles west of Okinawa, lies the small oval-shaped island Ie Shima. The main geographical feature on this five-mile-long island is a rocky peak, which sharply jets 565 feet into the sky, Iegusugu (Mount Gusuku) called by our soldiers, " The Pinnacle."

The location of Ie Shima was desired by the United States as an airbase that could be used to launch future air attacks against Japan using the B29 Superfortress bomber.

While still in the flotilla near the Kerama Islands orders arrived on the 11th of April to assault Ie Shima on the 16th, of April.

On the 15th of April, an artillery battalion was placed on a sandbar about 6,000 yards southeast of the island. On the morning of the 16th, heavy bombardment of the island began from several naval ships, plus artillery from our soldiers on the sandbar. Soon the 305th RCT and the 306th RCT arrived in an amphibious assault from the southwest. Once on the beach, enemy resistance was light. Cautiously approaching the plateau they soon discovered the island was covered with landmines. Steadily moving forward the combat teams soon pushed into very heavy enemy fire as they approached the town and the Pinnacle.

On the afternoon of April 17th, in a storm of enemy firepower, I was part of two battalions ofl the 307th RCT that came ashore from the south, near Ie Town. Advancing into the town we found the enemy had fortified positions. (Months in advance of our arrival the enemy had constructed pillboxes in houses, concealed machine guns behind concrete tombstones, placed weapons in tunnels under the town, and created stealth mortar positions; rigged explosive booby traps attached to drums of gasoline and placed landmines in normal paths. The most fortified area was a ridge overlooking the town on the Pinnacle.) There were also hundreds of local citizens armed and firing from houses. Once again in this intense battle, we encountered fanatical suicidal Banzai counterattacks. Families even used their children as human bombs carrying satchel charges. At times our soldiers were attacked by mobs with machine guns, bayonets, clubs and grenades. This battle lasted day and night for five days. The most intense part of the battle was on a ridge of the Pinnacle which was almost impregnable, we called it Bloody Ridge because of the many soldiers we lost during the fight. At Bloody Ridge, our commanders called for reinforcements several times. Finally, our Army surrounded the island and secured the Pinnacle on the 21st , of April. In the end, all that was left on this island was rubble. The 307th RCT left the island on the 25th of April, our next mission- Okinawa.

Unfortunately one of the losses in this battle was famous war correspondent Ernie Pyle.

*For us, the battle cost the 77th Infantry 239 lives, 897 wounded and 19 missing.

The Japanese losses were 4,794 killed and 149 prisoners.

Soon after the battle of Ie Shima construction of a B29 Superfortress Bomber base was started.

*Numbers from 77th Infantry Division Unit History page 251

The flowing are excerpts from a few documents showing some of the awards, and medals Sergeant Rathel was awarded during the Pacific Campaign during World War II.

MILITARY HISTORY

22. DATE OF INDUCTION	23. DATE OF ENLISTMENT	24. DATE OF ENTRY INTO ACTIVE SERVICE	25. PLACE OF ENTRY INTO SERVICE
21 OCT 42		4 NOV 42	FT MC CLELLAN
SELECTIVE SERVICE DATA — 26. REGISTERED: YES X — 27. LOCAL S.S. BOARD NO. # 3		28. COUNTY AND STATE: DOTHAN ALA	29. HOME ADDRESS AT TIME OF ENTRY INTO SERVICE: SEE 9
30. MILITARY OCCUPATIONAL SPECIALTY AND NO.: SECTION LEADER 604		31. MILITARY QUALIFICATION AND DATE (i.e., infantry, aviation and ...): CTB; MKM R NOV 43;	

32. BATTLES AND CAMPAIGNS

SOUTHERN PHILIPPINES, WESTERN PACIFIC, RYUKYUS,

33. DECORATIONS AND CITATIONS

ATO MED; APTO MED; PHILIPPINE LIBERATION RIBBON WITH ONE BRO BRONZE INDIAN ARROWHEAD; WORLD WAR II VICTORY MED; GOOD COND

34. WOUNDS RECEIVED IN ACTION

NONE

LATEST IMMUNIZATION DATES

1. By direction of the President, under the provisions Executive Order 9419, 4 February 1944, (Sec II, WD Bul. 3, 19 a Bronze Star Medal is awarded to Sergeant (then Private) Clarence Rathel, RA 34 395 943, 307th Infantry Regiment, for exemplary conduct in ground combat against the armed enemy du the Western Pacific Campaign in the Pacific Theater of Operat

Okinawa is the largest island in the Ryukyus Islands

Sergeant Rathel's Bronze Star Medal

The Bronze Star is a military medal awarded for heroic achievement, heroic service, meritorious achievement, or meritorious service in a combat zone

April 12, 1945

The Death of Our President and Commander in Chief, Franklin Delano Roosevelt

While the 77th Infantry Division prepared for The Battle of Ie Shima, President Franklin Delano Roosevelt passed away on the 12th of April 1945 in Warm Springs, Georgia. We heard the news of his death after the Battle of Ie Shima while preparing for The Battle of Okinawa. He had been the President of the United States of America since 1933 and was loved by our military and Nation because of his excellent leadership. Our Vice President Harry S. Truman then became the 33rd President of the United States of America.

32nd President of The United States of America Franklin Delano Roosevelt, and First Lady Anna Eleanor Roosevelt

*Picture The Museum of Columbus, Columbus, Georgia

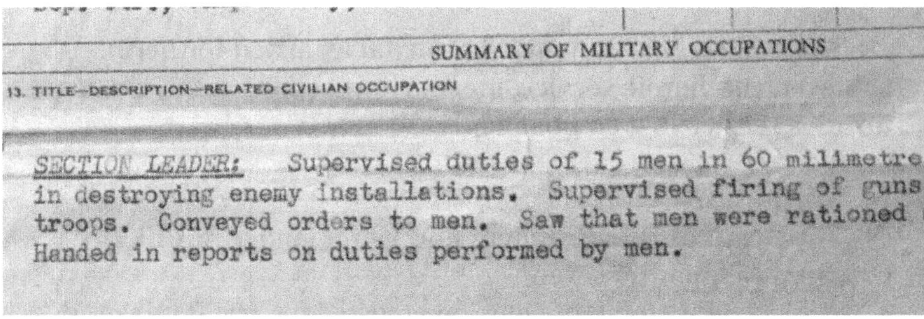

SUMMARY OF MILITARY OCCUPATIONS

13. TITLE—DESCRIPTION—RELATED CIVILIAN OCCUPATION

SECTION LEADER: Supervised duties of 15 men in 60 milimetre in destroying enemy installations. Supervised firing of guns troops. Conveyed orders to men. Saw that men were rationed Handed in reports on duties performed by men.

Sergeant Rathel was over a 60mm mortar unit during the Land Battle of Okinawa

It was said by others that he cared more about the men in his platoon than he did for himself.

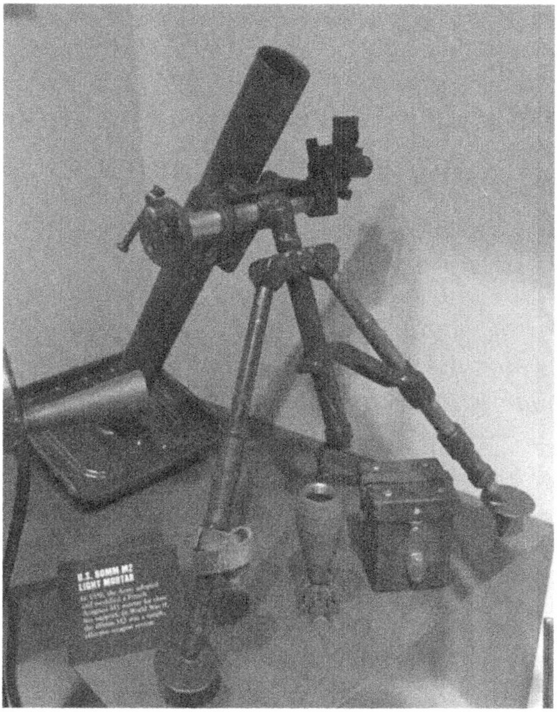

Picture of 60mm Mortar- The National Infantry Museum, Fort Benning, Georgia

***

Chapter Sixteen

The Land Battle of Okinawa – The Bloodiest Battle in the Pacific Campaign

Overview

The Land Battle of Okinawa was from April 1 to June 22, 1945, Codename of the Battle of Okinawa was ***Operation Iceberg****. The Allies planned to defeat the Imperial Japanese on Okinawa and from there attack Japan in a battle codenamed* ***Operation Downfall****.*

The distance from Okinawa to Japan's main island is 400 miles. The island of Okinawa is approximately sixty miles long and sixteen miles wide at its widest point, its climate is subtropical and during the spring its daytime temperature averages between 70 and 80 degrees. (This was a springtime war.)

The U.S. Army and Marine Corps invaded Okinawa on April 1st, 1945, in the largest Amphibious Assault in the Pacific Theater of War during World War II.

The 77th Infantry Division was part of the ***XXIV Corps.*** *The 7th, 27th, 77th, and 96th, Army Infantry Divisions combined with the Philippine Commonwealth Army and the Philippine Constabulary Army to form the* ***XXIV Corps. The XXIV Corps and the III Amphibious Marine Corps of the U.S. Marine Corps combined to form the Tenth Army.*** *The Tenth Army had its own Tactical Air Force and the Navy's Fifth Fleet.*

The estimated manpower for the Allies shifts between 539,000 to 541,000 depending upon the timeline of the war.)

The Imperial Japanese, Thirty-second Army, and Naval Fleet included their Air support. The estimated manpower for the Imperial Japanese shifts between 115,000 and 116, 000, ***in the Battle of Okinawa, true figures are a guess because the Imperial Japanese conscripted the citizens of Okinawa, including school children.***

The advantage the Imperial Japanese had in this battle was their unique fortifications on escarpments, ridges, and hills; their use of fortified tunnels, allowed weapons to be hidden in caves.

They also used self-sacrifice as offensive and defensive weapons.

Sergeant Rathel, Speaking

The 307th RCT, (Regimental Combat Team) was on its way to the island of Okinawa, on the 15th of April when our orders were changed to go to Ie Shima. **In the battle of Ie Shima which lasted just a few days, we lost almost as many good soldiers as we did in Guam.** I'm sure Ie Shima will go down in history as one of the hardest and bloodiest battles fought in World War II.

As we departed Ie Shima on the 25th of April, for the Battle of Okinawa, I noticed many of the faces in my regiment were missing.

***

As we sailed toward another battle deep within my soul, I wondered would the sacrifice of our fallen soldiers be remembered with Honor.
Looking back at this moment in time, I must say, we didn't realize that we would soon be in the battle of all battles, "The Battle of Chocolate Drop Hill."
To be Continued............

24. TOTAL NET SERVICE COMPLETED FOR PAY PURPOSES
27. DECORATIONS, MEDALS, BADGES, COMMENDATIONS, CITATIONS AND CAMPAIGN RIBBONS AWARDED OR AUTHORIZED
Combat Inf Badge With Sivler Star, Korean Svc Med W/5 Bronze Ca
ROK Pres Unit Citation, UN Svc Medal
28. MOST SIGNIFICANT DUTY ASSIGNMENT
29. WOUNDS RECEIVED AS A RESULT OF ACTION WITH ENEMY FORCES (Plac

The Battle Of The Chosin Reservoir

Made in the USA
Monee, IL
13 September 2026

62976628R00056